MULTI-LEVEL SIMULATION FOR VLSI DESIGN

THE KLUWER INTERNATIONAL SERIES
IN ENGINEERING AND COMPUTER SCIENCE

VLSI, COMPUTER ARCHITECTURE AND
DIGITAL SIGNAL PROCESSING

Consulting Editor

Jonathan Allen

Other books in the series:

Logic Minimization Algorithms for VLSI Synthesis. R.K. Brayton,
  G.D. Hachtel, C.T. McMullen, and A.L. Sangiovanni-Vincentelli.
  ISBN 0-89838-164-9.

Adaptive Filters: Structures, Algorithms, and Applications.
  M.L. Honig and D.G. Messerschmitt. ISBN 0-89838-163-0.

Computer-Aided Design and VLSI Device Development. K.M. Cham,
  S.-Y. Oh, D. Chin, and J.L. Moll. ISBN 0-89838-204-1.

*Introduction to VLSI Silicon Devices: Physics, Technology and
  Characterization.* B. El-Kareh and R.J. Bombard.
  ISBN 0-89838-210-6.

Latchup in CMOS Technology: The Problem and Its Cure.
  R.R. Troutman. ISBN 0-89838-215-7.

Digital CMOS Circuit Design.
  M. Annaratone.   ISBN 0-89838-224-6

The Bounding Approach to VLSI Circuit Simulation.
  C.A. Zukowski.   ISBN 0-89838-176-2

MULTI-LEVEL SIMULATION FOR VLSI DESIGN

by

Dwight D. Hill
AT&T Bell Laboratories

David R. Coelho
ZetaTech, Inc.

KLUWER ACADEMIC PUBLISHERS
Boston / Dordrecht / Lancaster

Distributors for North America:
Kluwer Academic Publishers
101 Philip Drive
Assinippi Park
Norwell, Massachusetts 02061, USA

Distributors for the UK and Ireland:
Kluwer Academic Publishers
MTP Press Limited
Falcon House, Queen Square
Lancaster LA1 1RN, UNITED KINGDOM

Distributors for all other countries:
Kluwer Academic Publishers Group
Distribution Centre
Post Office Box 322
3300 AH Dordrecht, THE NETHERLANDS

Library of Congress Cataloging-in-Publication Data

Hill, Dwight D.
  Multi-level simulation for VLSI design.

  (The Kluwer international series in engineering and
computer science. VLSI, computer architecture, and
digital signal processing)
  Bibliography: p.
  Includes index.
  1. Integrated circuits—Very large scale integration—
Design and construction—Data processing. 2. Computer-
aided design. I. Coelho, David R. II. Title.
III. Series.
TK7874.H525 1986   621.395   86-20869
ISBN 0-89838-184-3

Contents

List of Figures

Acknowledgments

I (Hill) would like to thank Willem vanCleemput for serving as my thesis advisor while working on ADLIB/SABLE. Without his guidance, and careful pruning of ideas, ADLIB would never have existed. I would also like to thank AT&T Bell Labs, Inc., for allowing me to work on this book and use company facilities to edit and print it.

I (Coelho) would like to thank Chalapathy Neti, Rick Lazansky, and Warren Cory for the work they did on HHDL and HELIX, and their contributions to timing verification and symbolic simulation, which constitute a large part of Chapter 5. I would also like to thank Silvar-Lisco, Inc., for the opportunity to work on HHDL and HELIX, and for their support of this book.

MULTI-LEVEL SIMULATION FOR VLSI DESIGN

CHAPTER 1

INTRODUCTION
AND
BACKGROUND

1.1 CAD, Specification and Simulation

Computer Aided Design (CAD) is today a widely used expression referring to the study of ways in which computers can be used to expedite the design process. This can include the design of physical systems, architectural environments, manufacturing processes, and many other areas. This book concentrates on one area of CAD: the design of computer systems. Within this area, it focusses on just two aspects of computer design, the specification and the simulation of digital systems. VLSI design requires support in many other CAD areas, including automatic layout, IC fabrication analysis, test generation, and others. The problem of specification is unique, however, in that it is often the first one encountered in large chip designs, and one that is unlikely ever to be completely automated. This is true because until a design's objectives are specified in a machine-readable form, there is no way for other CAD tools to verify that the target system meets them. And unless the specifications can be simulated, it is unlikely that designers will have confidence in them, since specifications are potentially erroneous themselves. (In this context the term *target system* refers to the hardware and/or software that will ultimately be fabricated.)

On the other hand, since the functionality of a VLSI chip is ultimately determined by its layout geometry, one might question the need for CAD tools that work with areas other than layout. Indeed, many non-trivial chips have been built by designers working at the "polygon level." The problem with this *bottom-up* approach is that it works only in limited contexts. When there is a clearly defined function of the chip, such as a random access memory (RAM), and the designers are familiar with the problem (perhaps they have done a similar one recently) and there are only a few designers working at one time (perhaps just one), then it may be possible to skip the specification/simulation stage and proceed directly to layout. This approach can be deadly, however, if one designer's understanding is subtly different from another's; their chips (or parts of a single chip), may not work together Furthermore, if the specifications change, the whole layout may have to be scrapped, since the relationship between any portion of it and the whole may be poorly understood. For these reasons, layout-based methodologies are today considered unreliable and unmanageable at even moderate scales of integration, and as VLSI systems become more complex, the importance of specification and simulation increases [COP74]. Even a designer who is able to understand completely an MSI chip with 500 transistors is likely to have a hard time with a VLSI chip with 50,000 transistors.

To avoid such problems, CAD researchers invented specification languages, often called Hardware Description Languages (HDL's), to help the designer communicate what is to be built, and secondly, how it is to be built. High-quality communication is essential within teams of designers. Communication is also required throughout the life-cycle of a project, *e.g.* from designers at early stages of a project (such as architects) to those at later stages (such as logic designers). Whereas a good specification environment can also help each individual designer to think more clearly about the target system and its environment, a poor or ambiguous medium can undermine the most skilled worker. Thus it is critical that the specification tools be powerful and precise; any concept that they can not express cannot be processed or verified in the CAD environment.

Once a specification is available, simulation tools can help confirm that the implementation matches the intended behavior. Classical simulation is not able to prove the absence of all errors except by *perfect induction* (exhaustively trying all possibilities), which is impractical in all but a few trivial cases. However, simulation can still be an effective engineering technique since simulation, using a well-made set of tests, can often identify many of the most frequently occurring types of errors, and do so while correcting them is still relatively cheap [BAR77-1]. Furthermore, in some cases *symbolic simulation* is able to verify correctness in an abstract and analytical way. (Symbolic simulation using ADLIB is discussed in Chapter 5.) To do either type of simulation, however, one must have a good way to drive the simulator and judge its output. There are two schools of thought on this: *functional testing* and *implementation-based testing*. Purely functional tests are independent of the implementation. For example, a multiplier chip simulation might be fed a few numbers, and if it produced the correct answers, it would be assumed to work. No knowledge of its internals would be required. Implementation-based testing, on the other hand, requires a detailed knowledge of the implementation and the most likely ways it could fail. For example, if it were known that the multiplier had a ripple-carry chain that was the speed-determining part of the circuit, an implementation-based set of tests might try to find an input combination that exercised that path. Test sets designed to catch VLSI fabrication errors are, by their nature, in the second category, since they must model the types of faults that can arise in one particular implementation. In general, the problem of creating a good set of tests is difficult, since it must be done with a thorough understanding of the intended behavior of the system and the most likely ways that the implementation can fall short. An unambiguous specification of both intention and implementation is therefore a prerequisite for writing any useful set of tests.

1.2 Categories of Simulation

Because simulators are so useful, many have been developed. Although each can be effective in its own domain, unfortunately, no one simulator is useful throughout all phases of design. Instead, at least three types are normally used as a new system is developed.

1.3 Performance-Analysis

First, high-level resource issues may be analyzed using a simulator such as GPSS [SCH74], PAW [MEL85], or SSH [ROH77]. The concepts modeled at this level are often statistical, such as bus contention as a percentage of bus accesses, the frequency of buffers overflowing, and so forth. This type of modeling helps determine the key architectural parameters, such as the number of registers and required bus bandwidth, in light of an estimated workload. It also predicts performance. If performance is not adequate at this level, the architecture and/or the specifications must be reworked or renegotiated, since subsequent design steps can do nothing to improve performance, and will probably degrade it as technology constraints are added.

1.4 Register-Level Simulation

Once analysis at the statistical level is completed, or sometimes in parallel with it, the design may be refined into more concrete hardware and software components. At this point the target system may be redescribed in an intermediate-level language such as the "Instruction Set Processor" language (ISP)* [BEL71]. This level of simulation is intended to discover anomalies in the instruction set or its interpretation. For example, two operands of a complex instruction can read and modify the same base register and interfere with each other destructively. This level of specification is often called the *principles of operation* for a VLSI chip. It does not specify how the chip is to be implemented, but it is usually specific enough to judge whether a given chip is working in a reasonable and useful way.

Some of the early languages that operated near the ISP level include APL [FAD64], AHPL (A Hardware Programming Language) [HIL74], Computer Design Language (CDL)[CHU65] [CHU74], and Digital Design Language (DDL) [DIE75] [COR79]. Typically, languages at this level provide extensive bit-manipulation facilities and allow powerful, concise specification of many register-oriented computer activities. CDL was one of the first hardware description languages. The basic ideas of CDL are similar to the more recently developed Guarded Command Language [DIJ75] by Dijkstra. A CDL program consists of a set of statements, each with a special "label" or "guard" that determines when it should be executed. ALGOL control constructs are allowed within statements, and synchronous clocks of several phases may be used. DDL also provides bit manipulation facilities, but adds the concept of finite state machines (FSM's) as its basic timing and control mechanism. This comes naturally to many machine designers, who can rely on a large body of theory devoted to FSM's. Because both the guarded parallel processes and the FSM facilities are natural and important abstractions, both will be discussed in a later chapter. Many "silicon compiler" projects have been striving to convert descriptions from this level into hardware automatically.

1.5 Functional-Level Simulation

On the other hand, there are systems that use general-purpose programming languages for machine descriptions. Many gate-level simulators allow the user to insert arbitrary code to describe the function of components. One of the earliest was SALOGS[CAS79], which uses routines written in FORTRAN; dozens of other simulators have been

---

* ISP was originally developed just for describing computer instruction sets in a textbook. Since ISP was only a printed language it was soon replaced by ISPL. and later ISPS [BAR79]. which are improved versions suitable for computer parsing and interpretation.

built that use several other languages. For example, BUILD [LAN79], which was designed at Burroughs Corporation, provides facilities for describing a variety of register and gate activities. It makes use of the bit manipulation facilities in the Burroughs extensions to ALGOL. Other programming languages that have been used or considered for hardware description include SIMULA67 [DAH68], Concurrent Pascal [HAN77] and MODULA [WIR78], C, and Concurrent C[GEH85]. These are modular, block-structured languages that can readily describe software components that interact via shared variables and monitors. Unfortunately, this is not a good model for the way that hardware components interact, and the task of designs in these languages gets progressively more difficult as the design is refined to more detailed hardware constructs.

1.6 Gate-Level Simulation

Once the ISP level work is complete, designers can progress to a more detailed register or gate-level description, which is then described and simulated in another language such as TEGAS [SZY73], or F/LOGIC [WIL76]. Here the principles of operation document is the guide for designers, who assign operations to specific physical blocks such as adders, busses and NOR gates. In custom MOS VLSI design, work at this level often concentrates on well established design techniques such as PLA's, standard cells, or *gate-array* layout techniques. In current practice this level of simulation often requires a great amount of computing resources since a large fraction of the target machine is represented at a detailed level.

Discussion: Hardware Accelerators

To speed up simulation, special-purpose hardware accelerators may be used. These come in two varieties. The first is a mapping of a low-level software simulation into hardware: each gate in the target machine is represented by a specific point in the simulation engine's memory. These can achieve a speed improvement proportional to the ratio of hardware speed to software speed, multiplied by some degree of parallelism, (which is at best a constant factor for any given hardware accelerator). Unfortunately, such accelerators cannot always take advantage of special knowledge of the target system that may be available to software simulators. Also, they cannot be used until a detailed model of the logic is available, so their role is often one of "certification" of nearly complete designs.

The second type of hardware accelerator consists of more conventional computers augmented with a special socket, into which the user plugs actual VLSI chips. This allows the logical behavior of that chip to be specified by the chip itself, which has two benefits. First, it may be orders of magnitude faster than a software simulation of the chip*. Secondly, it may allow a chip that is not completely specified to be included in the simulation. This is especially important when dealing with chips coming from other vendors, since they might not come with complete and accurate documentation. (Even the vendor may not have all the detailed specifications.) On the other hand, this method has obvious limitations: if the chip has not yet been fabricated, it cannot be simulated. Furthermore, it does not provide any insight into the internals of the chip in the socket, just its external behavior. Finally, there is a danger that the external behavior of one particular part from one particular vendor may not represent precisely the intended behavior of that class of chips. The absence of complete and comprehensive specifications can lead to anomalous behavior later when other parts from other vendors are substituted.

1.7 Analog Simulation

If the design includes analog circuitry, or high-performance digital circuitry, yet another simulator, such as Mini-MSINC [ANT78], or SPICE [NAG73] may be necessary for circuit analysis. Although one might consider simulating an entire machine at the circuit level, this is seldomly done because the computing resources required would be enormous. Rather, the designer first identifies any paths that are critical from a timing or electrical standpoint, isolates them, and then simulates them with specially derived input data. Even so, circuit simulation is today a major consumer of high-end computer resources. Midway between gate-level simulators and circuit simulators are *switch-level* simulators such as MOSSIM[BRY81], SOISIM[SZY84], or LOGMOS[DUN83]. These represent transistors as ideal switches, and resolve the behavior of many MOS circuits that could not be efficiently simulated any other way, such as circuits using "pass transistors." Another set of tools that bridge the gap between logic and circuit simulation, but in a different way, are those characterized by SPLICE [NEW78], mixed-mode MOTIS[AGR80], and DIANA[DEM79]. Each of these combine an efficient circuit

---

*   But not necessarily - some high level ADLIB models have been shown to run faster than some hardware accelerators

simulator with a logic simulator, so that the circuit simulation can be done in the context of the logic gates that are connected to it.

1.8 Problems with the Existing Approach

The design process outlined above consists of multiple specification/ simulation cycles. In each cycle the designer is progressing from a more abstract to a more concrete form of the design, and is binding design decisions along the way. Although the earlier design decisions are in some sense the most critical because they impact all subsequent stages, in practice all decisions are important. This is because VLSI circuits are notoriously "unforgiving," and a mistake at any stage is usually fatal, and may bring with it major negative consequences, including the inability to test the remainder of the chip, an additional delay of as much as several months (even for minor fixes), and thousands of dollars of rework costs. These consequences are perhaps the most compelling reason why simulation of hardware has assumed such an important role.

The multiple simulator approach outlined above has two advantages and many disadvantages. The advantages are that each simulation can be written in a language tuned for one particular level, and that each simulator can optimize its run-time organization for one particular task. The disadvantages include the following:

1. Design effort is multiplied by the need to learn several simulator systems and recoding a design in each. User education and motivation becomes a problem.

2. The possibility of error is increased as more human manipulation is required.

3. Each simulator operates at just one level of abstraction. Because it is difficult to simulate an entire computer at a low level of abstraction, only fragments can be simulated at any one time.

4. Each fragment needs to be driven by a supply of realistic data and its output needs to be interpreted. Often, writing the set of tests to serve these needs is as much effort as developing the design itself, since it requires a complete knowledge of the specifications and the design augmented with a detailed and thoughtful understanding of the things most likely to go wrong.

5. As the design becomes increasingly fragmented in order to simulate it, it becomes difficult for any designer to see how his or her piece fits into the overall function of the system.

6. Comparison and validation of simulation results across different levels becomes next to impossible when they are generated from

different tools.

To avoid the problems listed above, a simulation environment is needed that can be used from the beginning to the end of the design process. It should enable the designer to use behavior specification and simulation not as a separate operation, but as an integral part of the design process. Ideally, simulation should require no additional effort by the designer, but should provide useful information about design options as soon as possible. The most important information, of course, is identifying errors.

The simulation system should accept specifications in a way that is natural for the engineer and compatible with other design automation or fabrication systems. In addition, the simulator should be able to adapt to the individual using it, not the other way around. In particular, since there exist many distinct schools of thought in digital designs (such as PLA's, random logic, microprocessors, *etc.*), the simulator should be flexible in its approach and easily extensible by the user. Because the human mind tends to switch among several abstraction levels during the design process, the simulator system should support multiple levels of abstraction, and should allow several abstraction levels to be mixed together in one simulation. For example, when the designer finishes the refined design of one subsystem it should be possible to simulate it in the context of more abstract models of the remainder of the system. This allows him or her to ignore the details of the other components but at the same time use them to supply test data to the new design and analyze its output.

1.9 Multi-Level Simulation

A number of CAD specialists have recognized the utility of simulation tools that can operate at more than one level of resolution. As mentioned above, SPLICE, and DIANA operate at both circuit and gate levels. Intermediate between gate and chip levels are FDL [CHA76], MACSIM-PVS [TOK78], MULTI-SIM [ABR77], and the functional-level modeling discussed in [MAC75]. While DDL is a somewhat more abstract language and is more suitable for algorithmic descriptions, it is also possible to use DDL to describe individual gates.

The above tools are primarily aimed at closely related levels of hardware design, but there is also tools that span a wider ranges. MOTIS supports register-transfer level (FDL2), C language, unit delay, multiple delay and timing simulation. It also allows mixing of different levels in the same simulation. However, it does not have extensive facilities for very high level work, such as software modelling or protocol debugging. One tool that does started out as LOGOS [ROS75], and then evolved into System's

Architect Apprentice (SARA) [GAR77]. SARA provides a language called SL1 for defining a structured hierarchy of modules, their interconnections and interface sockets. The idea is to allow refinement of a module without affecting its neighbors as long as its sockets are not changed. SARA also contains a second tool which allows a user to specify a system's behavior and to simulate it. The behavioral information is entered as Graph Models of Behavior (GMB), one for control flow and one for data flow. The control flow graph resemble a Petri net with nodes, arcs and tokens. The data flow net contains processors, data sets and data arcs. It's activities are directed by the control graph on a node-by-node basis.

Advantages of the SARA system include the ability to manipulate topology and nesting via SL1, the use of analytical techniques to detect races and ambiguities in the GMB, and the availability of a general-purpose language (PL/I) for describing the behavior of data processing nodes. Disadvantages include the need for a user to enter multiple representations of a single system, and the difficulty of splitting behavior into separate control and data functions.

While each of the above languages has merits, none of them satisfies all the needs of a comprehensive design specification and simulation system. In particular, those that work well for modeling hardware structures provide little support for modeling software, which is important at the more abstract levels of design, and *vice versa*. Therefore, it was decided to begin work on a new language. But since a completely new language would not take advantage of the programming expertise possessed by many of today's designers, it was decided to start with a popular and generally available language, and add as few features as possible, while retaining the syntax and semantics of the base language.

1.10 Genesis of ADLIB

A first attempt at multi-level simulation was based on SIMULA67, since this language was designed for both simulation and extensibility. The result was the "ALGOL Derivative Language for Indicating Behavior" (ADLIB). Although it proved to be adequate to build a working simulator, there were several problems with it. First, SIMULA67 lacked effective type-definition and type-checking facilities, which are essential for the early detection of errors*. Another drawback was that

---

* SIMULA67 does have an interesting CLASS definition facility, but this proved a poor fit for describing some of more detailed properties of hardware.

SIMULA67's high level multi-tasking facilities required a significant amount of run-time overhead, (though a globally optimizing compiler might reduce this). However, the biggest difficulty was the size of the language. SIMULA67 is a large language by itself, and with the addition of special features for CAD it became difficult to manage.

It therefore was decided to restart the effort based on Pascal [JEN74]. Pascal is a smaller language than SIMULA67, and at the same time it provides a powerful mechanism for the user to declare new data types. This is in keeping with the design principles of low overhead and simplicity for the user. The resulting language was again named ADLIB; which this time stood for "A Design Language for Indicating Behavior." ADLIB and its support package, the "Structure and Behavior Linking Environment" (SABLE), were designed and implemented at Stanford University by Hill in 1979, as his Ph.D. thesis*. A second implementation of ADLIB, with a number of new language features and an "industrial strength" support environment, was done under the direction of Coelho in 1984**. The new system was called HELIX, and the hardware description language was called Hierarchical Hardware Description Language (HHDL) [COE83]. Since HHDL is very similar to ADLIB the syntax and examples in this book are common to both languages, except as noted in the Appendix. The primary enhancement provided by HHDL is the *package* and *module* constructs, that allows groups of related objects to be grouped together, compiled separately, and maintained as a unit. The support environments differ more than the languages. While both environments perform the task of joining the behavior descriptions of ADLIB with a structure description, considerably more facilities for this are present in HELIX than in SABLE, primarily dictated by the requirements of practical industrial designs. Since these facilities are well documented in the HELIX manual, this book will not discuss the support environment in any detail, except as it relates to the language itself.

1.11 Outline of This Book

Chapter 2 explains the notion of multi-level simulation as supported by ADLIB, but does not go into specific details. These are explained in Chapter 3, which gives the syntax and semantics of each ADLIB construct. Chapter 4 builds on Chapter 3 by illustrating how the

---

* Copies of this implementation, which are tuned to the DEC-10 Pascal, are available from the Stanford University Office of Technology Licensing.

** Available from Silvar-Lisco Incorporated, Menlo Park, CA.

constructs fit together to describe typical hardware facilities, mostly from a hardware designer's standpoint. Chapter 5 provides more complex examples using these facilities, and discusses the details of the language at a level appropriate for CAD workers. Since some knowledge of the implementation techniques is useful both to understand the language concepts, and to develop new simulators or extend old ones, Chapter 6 includes some of the basic techniques involved in efficiently implementing ADLIB. Chapter 7 considers the possible extensions and options that may impact ADLIB in the future, and sums up the key ideas.

CHAPTER 2

PRINCIPLES OF
MULTI-LEVEL SIMULATION

2.1 Structure and Behavior

The task of specifying and simulating a wide range of systems over the
complete design cycle is an inappropriate one for a single computer
language. This is because there are two fundamentally different kinds of
information that need to be expressed. The first is structure, which is
declarative in nature. That is, structure can naturally be expressed by a
series of declarations with little or no context or order dependencies.
(*E.g.* J touches B, B touches C, F touches C, *etc*.). The other kind of
information is behavioral, which is basically imperative. That is, each
behavioral operation has side effects that change the world around it.
Although there have been many attempts at describing behavior in a
declarative way, they have been only partially successful. In particular,
when describing complex behavior, an ordering of the statements
becomes important: the language breaks down into describing a set of
objects declaratively, where each of these objects is itself imperative.
For example, PROLOG is sometimes described as a declarative
language, but non-trivial PROLOG programs depend heavily on the
order of evaluation, not just to enhance performance, but for their
semantics as well [HIL85]. Conversely, some systems, such as VHDL,
include structure information within the behavior description language.
While this can have the advantage of a uniform syntax for the two
aspects of the design, their is usually little in common between their
semantics. The result is really two languages in one wrapper.

While one could argue indefinitely whether this need always be true, it seems more productive to accept this schism and make use of it. That is the approach taken here. Behavior is captured in "A Design Language for Indicating Behavior" (ADLIB), and structure is specified in the "Structural Description Language" (SDL)[VCWM77]. They are unified in the Structure and Behavior Linking Environment (SABLE), or under the Hierarchical Environment and Language for Interactive Circuit Simulation (HELIX). The simulator analyzes the structure of the target system and connects the appropriate behavioral descriptions accordingly.

2.2 Two Concepts of "Level"

To understand the use of ADLIB, it is first necessary to understand the way that it models the target system and how a user expresses his or her ideas at various levels. This notion of *level*, which has been used loosely so far, has two specific meanings. When the structure of a system is described as a hierarchy of components, each component is seen to occupy a specific *nesting level* in the hierarchy. This is the number of levels of refinement between the topmost (system) level and the component's own level. In SABLE/HELIX this information is conveyed in SDL.

On the other hand, when specifying the behavior of any particular type of component, one is concerned with a different notion of level, one that is determined by the data structures that the component manipulates, and its interpretation of those structures. For example, while a highly abstracted model of a virtual memory system might have an integer variable called pages_in_use that records the number of occupied disk pages, a lower level model may deal with a large array of bits that represents the data on those pages. This leads to the concept of *data level* as used in ADLIB: if one data structure and its interpretation serves as a abstract approximation of another, then the first is said to be at higher data level than the second. For example, an integer variable representing an address could also be represented by an array of real numbers representing the instantaneous voltages on a set of address lines. The integer address is a compact approximation of those voltages adequate for high-level simulation. This concept of data levels does not form a total ordering on the set of data representations since many levels may have no meaningful comparison. In general however, the nesting level of a component tends to be correlated with the data level of the structures that it manipulates. It should also be understood that an ADLIB model does not depend on any ordering of data levels. The ordering exists only in the user's mind (if at all) and may be expressed only through comments.

One important difference between these two concepts is that while a component must occupy exactly one nesting level structurally, a description of its behavior may manipulate information at any number of data levels. This is the key to simulating several levels of abstraction simultaneously.

2.3 Design Methodologies

A design methodology is a set of rules for a designer or design team to follow. The philosophy behind any methodology is that if a design is developed according to an orderly set of rules, the design process can be more efficient, and the resulting work itself will be more orderly, more correct, and more maintainable. The concept of components operating at various data levels is central to any top-down design methodology which encourages simulation prior to detailed design. To provide some background, it is helpful to first review the structural top-down methodology supported by tools such as SCALD [MCW78] that deal with multiple nesting levels but not multiple data levels.

Using one of these tools, a designer works at one nesting level at a time, defining the target system to consist of a group of components interconnected by nets. When this step is finished, he or she steps down one level and defines each of the components in terms of simpler components. This process is repeated recursively until the entire target system has been defined in terms of primitive building blocks. The macro hierarchy results in a highly structured system, but underneath, only the predefined primitives exist. This approach also implies that an entire subsystem must be completely specified in terms of basic blocks known to the simulator before simulation of it can begin. Therefore simulation is not available during the early stages of design when the detailed designs are not yet available. Another consequence of unrestricted expansion of hardware macros into primitive blocks is a uniform, low-level of abstraction, which can lead to overwhelming computation loads for the simulator. As a result, simulation can lag behind the current state of the design, and be unavailable to individual designers requiring quick and inexpensive feedback.

2.4 Top-Down Methodology Enhanced with Simulation

The ADLIB user also starts at the topmost nesting level of his or her design and decomposes it into a few interconnected components. However, before refining the structure of these components, the desired behavior of each component is specified by writing an ADLIB behavior for it. ADLIB is run, and the resulting simulation is used to evaluate overall configuration and performance. As the design progresses, the components are decomposed into smaller functional blocks, which may include both hardware and software units. The behavior of each is again specified in ADLIB. The decomposition and refinement of components is recursive, just as before. However, the user now has an objective criterion for evaluating the design at each point: if the simulated behavior of the refined design does not match that of the higher-level component then the refined design is presumed to be erroneous*. Obviously, it is important that the ADLIB models are written and tested carefully, since they become the definitive specifications for the next level of design.

At this point, an alternative methodology needs to be discussed. This is the middle-out technique, which is probably the one most commonly used. Whereas the top-down approach assumes a perfect understanding of the specifications and complete confidence in the ability to implement subsystems, the middle-out approach assumes neither. Rather, it starts with a rough idea of what the system is to do, and a rough idea of the available resources. The high-level requirements/specification data and the low-level implementation data are refined continuously. The process stops when 1) the requirements and specifications are acceptable to the environment and 2) the implementation is ready for fabrication and 3) they agree with each other. From the standpoint of tool development, the difference between the middle-out and top-down approaches is secondary, ADLIB can support either. A more significant difference lies with project management – top-down design tends to be more controllable but slower, since low-level work cannot begin until higher level work is completed, and multiple restarts may be needed when low-level modules prove impossible to design. By contrast, the middle-out methodology tends to be more flexible, but can lead to chaos, especially when there are multiple versions of requirements and implementations floating around. The decision of how to organize the workers in a large design effort is therefore a crucial one, and one that cannot be automated.

2.5 Example: Data Acquisition System

To illustrate briefly how one of these methodologies, the top-down approach could work, let us assume that a microprocessor data acquisition system is to be developed. The design has progressed to the point shown in Figure 2-1, where we deal with an array of two sensors called sensor1 and sensor2, a multiplexing unit called sense_mult,

---

*   In practice, both the design and the original specification get reexamined at this point.

and a bus interface called `bus port`. Before the design can proceed it is necessary to specify the intended behavior of each of these components. Component type `multiplexer` will describe the behavior of `sense_mult`, component type `bus_connect` will describe the behavior of `bus_port` and component type `sensor` will be used twice to describe the behavior of `sensor1` and `sensor2`. Concentrating on the multiplexer, we see that it must accept two eight bit data paths and multiplex them onto a single eight bit path, the choice to be determined by the state of the control net.

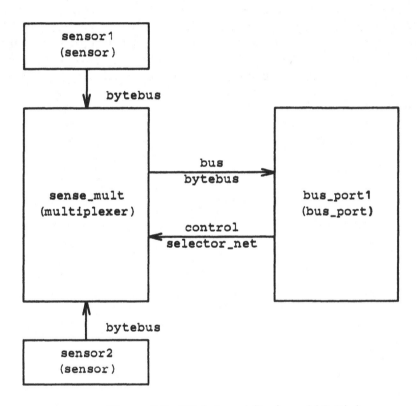

Figure 2-1. High-Level Design of Multiplexer

An ADLIB component type definition that describes the behavior of the multiplexer is shown in Figure 2-2. Component type definitions begin with the keyword COMPTYPE. (In the code examples throughout this book, all references to computer code are give in constant-width font to distinguish them. Furthermore, reserved words like WHILE are UPPER CASE, though the compilers are actually case-insensitive.) In addition, Figure 2-2 also contains the definition of three types of nets, introduced by the keyword NETTYPE. Net type bytebus is a set of eight bits, each of which may be accessed independently; net type selector_net indicates that either path1 or path2 is to be used. Net type boolnet is a more primitive net type that holds one bit of information. It will be used later in this section. The component type multiplexer declares that data1 and data2 are INWARD nets of net type bytebus, and that cntrl is an INWARD net of type selector_net. The component type also declares dataout to be an output net of type bytebus. Subprocess channel1 transmits the value of net data1 to net dataout, delaying the update for prop_delay time units. Subprocess channel2 does the same for data2. The main body of component type sense_mult determines which process will be permitted to operate. If net cntrl holds the value path1 then channel1 will run, otherwise channel2. In either case a WAITFOR statement is executed that causes execution of the main body to pause. The main body remains idle until some other component updates cntrl, at which point the cycle is repeated.

```
NETTYPE
   bytebus = set of 1 .. 8;
   selector_net = (path1, path2);
   boolnet = boolean;
COMPTYPE multiplexer;
   INWARD
      data1,data2 : bytebus;
      cntrl: selector_net;
   OUTWARD
      dataout : bytebus;
   SUBPROCESS
      chan1 : TRANSMIT data1 CHECK data1 TO dataout DELAY 5.0;
      chan2 : TRANSMIT data2 CHECK data2 TO dataout DELAY 5.0;
   BEGIN
      WHILE true DO
         BEGIN
         CASE cntrl OF
            path1:
               BEGIN
               permit(chan1);
               WAITFOR cntr1 <> path1 CHECK cntr1;
               inhibit(chan1);
               END;
            path2:
               BEGIN
               permit(chan2);
               WAITFOR cntr1 <> path2 CHECK cntr1;
               inhibit(chan2);
               END;
            END; (*case*)
         END;
   END. (*comptype multiplexer*)
```

Figure 2-2. Definitions of `Bytebus`, `Selector_net`, `Boolnet`, and `Multiplexer`

Using ADLIB to combine the structure of Figure 2-1 with the definitions of component types `multiplexer`, `sensor`, and `bus_interface` (not shown), the designer simulates until satisfied that each component, and the system as a whole, are behaving as intended.

2.6 Refinement of Design

Once this is completed, the design problem can be split into three independent parts at the next level of refinement, one for each of the three component types. Focusing on one of these, the multiplexer, our problem is to build a system of lower level components that will collectively function just like component type multiplexer. Suppose that it were decided to use two-way, one-bit multiplexers of component type bitm and connect them as shown in Figure 2-3. Each component accepts the net type boolnet which represents a single electrical connection.

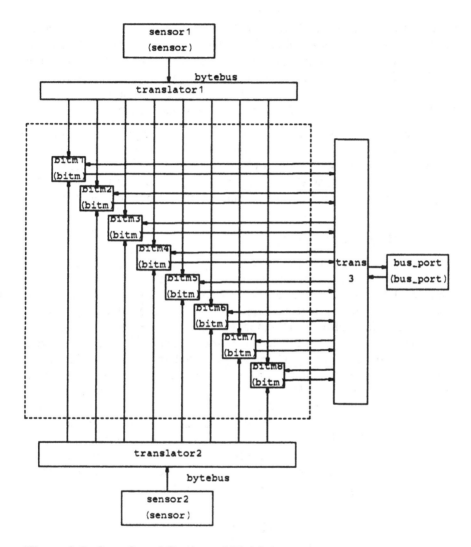

Figure 2-3. Low-Level Design of Multiplexer

To simulate the behavior of this system it is necessary to define another component type in ADLIB; this time it is bitm. But once this is done, the design can be validated in two ways: the overall system can be simulated to see if it behaves the same with the multiplexer subsystem in place of sense_mult in Figure 2-1. Alternatively, the isolated multiplexer can be compared with the results for the isolated sense_mult. The decision of which methodology to use depends on the degree of confidence one has in the testing process, each component's expected performance, and history.

If the system performs as before one can be reasonably confident in the refined design of the multiplexer, and bitm becomes the next target of our design efforts. The methodology is recursive, and uses simulation at each level *before* a component is decomposed into subcomponents. The design process terminates when the primitive component types are directly realizable, *i.e.* when they require no design themselves.

In the above discussion the question of how a single net of net type bytebus at one level suddenly becomes eight individual nets at a lower level was not addressed. This problem is actually solved in ADLIB by means of data-level translators, which will be discussed in Chapter 4. But first, it is necessary to expand on some of the principles touched on so far.

2.7 Stimulating A Component

Component type multiplexer from the previous section is an example of an ADLIB component that remains quiescent until stimulated into action by the update of one or more of the nets that it is connected to. This is the primary mechanism for controlling the behavior of components. In addition, a component may be defined that awakens itself after a delay. The use of input nets and self-awakening components are the only mechanisms provided for initiating the actions of a component; components may not activate each other directly. A component can react by altering its internal state and by assigning new values to its output nets. The correspondence between the stimulus a component receives on its input nets and the response it passes to its output nets completely specifies the behavior of a component, so that the actual coding or the state of its internal variables is irrelevant outside of its scope. The resulting modularity is also beneficial from a software engineering standpoint since it restricts side effects and tends to localize errors.

This stimulus - reaction discipline is not the only way to structure a simulator. Other approaches include: event driven systems such as SIMSCRIPT [KIV69] or ASPOL [LOS75]; unconstrained scheduling mechanisms as in SIMULA67; process-processor queues such as in those found in GPSS or SSH; and the GMB - Petri net approach used in SARA[GRI77]. It is usually not difficult to translate a given function from one system to another. However in designing ADLIB it was felt that the stimulus - reaction notion was best for several reasons. First, it is an extension of finite state machines, a concept familiar to most designers. Secondly, it is a general mechanism that can be applied to both software and hardware at highly abstract levels and at minutely detailed levels. For example, an operating system reacts to I/O requests by accepting or outputting data, and a transistor reacts to a change in its gate voltage by changing its impedance. Finally, its modularity makes this approach compatible with SDL [VCWM77] and the goal of multi-level simulation.

2.8 Importance of Nets

Because nets are the fundamental way in which components communicate, they are used to transmit both data and control. Any distinction between the two is up to the user and ADLIB does not enforce it. The user may define as many net types as desired, just as bytebus and selector_net were defined in Figure 2-2. Components may assign several nets to be updated at the same simulated instant, but if two components attempt to update the same net to different values in the same event, the behavior of the system is undefined. This implies that whenever wired-or is used, components must be added to perform this function*.

2.9 Summary

This chapter presented the basic model used by ADLIB to represent digital systems. This model is based on independent processes communicating over pathways called *nets*. The topology of these nets represents a one kind of design knowledge; it is specified in a structure describing language, which may be graphical or textual (SDL). In most designs, this knowledge starts off being very simple at highly abstract levels, and highly detailed as the design progresses and more structure is created. On the other hand, the actions of the components represents a different type of design knowledge. At highly abstract levels, these actions tend to be very complex because an entire design is described in just a few components. But later, when the objects being described represent low-level primitives, the actions become very simple. The semantics of ADLIB, which are the subject of the next chapter, are designed to make the job of describing this behavior as easy as possible at both high and low levels.

---

*   This is handled automatically in HELIX.

CHAPTER 3

THE ADLIB LANGUAGE

3.1 Purpose of ADLIB

The purpose of an ADLIB description is to define the intended behavior of one or more types of digital components. In the simulation environment these are combined with topology information that specifies the number of components used and the way they are connected. This topological information may be expressed graphically or with SDL.

The ADLIB construct that defines a component's behavior is called a component type definition, and is introduced by the keyword COMPTYPE. All of the syntax and sematics of ADLIB are designed to make it easier to write component type definitions that model hardware and software systems.

3.2 Pascal Control Constructs

Because ADLIB is a superset of Pascal it includes all of the Pascal control statements, which are summarized here.

1. The IF statement,

    IF <boolean expr> THEN <stmt1> ELSE <stmt2>

    which chooses between two alternative statements.

2. The CASE statement,

```
CASE <expr> OF
    <value1>:<stmt1> ;
    <value2>:<stmt2>;
    END
```

which selects one of an arbitrary number of statements (similar to, but more powerful than a "switch" or "computed goto").

3. The WHILE loop,

    WHILE <boolean expr> DO <stmt>

which iterates a statement zero or more times.

4. The REPEAT loop,

    REPEAT <stmt list> UNTIL <boolean expr>

which iterates a statement one or more times.

5. The FOR loop,

    FOR <variable>:=<expr1> (TO | DOWNTO)
        <expr2> DO <stmt>

which repetitively executes a statement as <variable> ranges from <expr1> to <expr2>. (Similar to a FORTRAN DO loop.)

6. The unconditional GOTO,

    GOTO <label>

which transfers control to <label> unconditionally.

The above six constructs are useful for defining the algorithm incorporated within a component, but are not adequate for describing inter-component control and data flow. Therefore, the following new constructs are included in ADLIB:

3.3 Additional ADLIB Constructs

1. The ASSIGN statement,

    ASSIGN <mode> <expr> TO <net name> <timing clause>

The ASSIGN statement evaluates <expr> and stores the result away in a hidden area. At a later time, this value is retrieved and assigned to the specified port. Time delays may be specified in several ways depending on the nature of the circuit (synchronous or asynchronous) and the objectives of its designer. The simplest way is to define a delay directly, *e.g.*:

```
ASSIGN true TO out DELAY 15;
```

Fifteen simulated time units after this statement is executed, the net connected to port out will be updated to the value true. Time delays need not be constants — any expression may be used. For example, if two parallel paths exist to the same outward port, and either one can drive it, then the component's behavior could be defined as:

```
ASSIGN result TO out DELAY min(delay_1,delay_2);
```

(Min is a function that returns the minimum of its arguments.)

The expression in an ASSIGN statement may contain function calls. For example, in order to describe a signal generator it may be convenient to write:

```
ASSIGN sin(time*frequency) TO signalout;
```

This statement illustrates two other points as well. First, it makes use of time. In ADLIB, the global read-only variable time always contains the current value of the simulation time. When the simulation begins, it is equal to 0. Secondly, it does not contain an explicit timing control clause. The ADLIB compiler therefore treats it as if DELAY 0 were specified. At first glance zero propagation delay times may seem confusing, unrealistic, and potentially hazardous. However, because of the way that ADLIB simulation is defined, this operation is unambiguous and in certain applications can be useful. The simulator cycles between the execution of components the updating of nets connecting them. First, all components are allowed to execute, then all nets are updated, then all components are allowed to execute again, *etc.* One iteration of this cycle constitutes one *event*. It may happen that several events occur sequentially, but at the same simulated time. If one or more components ASSIGN to a set of nets with a DELAY of 0, then all these updates will appear to occur simultaneously.

No hazards or races are introduced by allowing zero propagation delay, and there are several applications where it is necessary and appropriate. For example, a designer may prefer to treat combinational logic as operating with zero time delay to contrast it with sequential circuitry. An extreme example of this would be a system implemented with relays. A reasonable approximation might be to assume that a voltage propagates through the contacts of a relay immeasurably faster than the speed at which the armature moves. To describe a relay which operates as a single-pole, double-throw switch (like a one bit multiplexer) one could write:

```
IF armature_position = up THEN
        ASSIGN input1 TO out
ELSE
        ASSIGN input2 TO out;
```

In this example the exact speed of propagation is irrelevant. On the opposite extreme are synchronous circuits whose output values must be available at precisely controlled instants. For example, most micro-controllers operate at a constant speed independent of the micro-instruction mix (this is not always true of macro-instructions). Such controllers, and any circuitry directly connected with them, are most conveniently defined in ADLIB with the use of the CLOCK SYNC primitives. An ADLIB clock may be thought of as a function that maps simulation time into positive integers. At time 0 all clocks have value 0. As simulation time progresses, the clocks run through their phases repetitively: *i.e.* 0,1,2,3,0,1,2,3,0,1,2,3... for a four phase clock. The period of repetition is the parameter value specified by the user in the clock definition statement. The value of clock clk, defined as:

```
CLOCK clk(8,4);
```

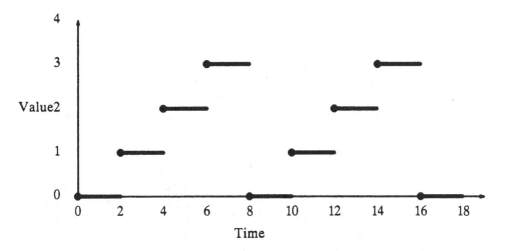

Figure 3-1. Value of Clock clk(8,4)

is shown in Figure 3-1.

The SYNC operator can be used to synchronize an operation with a particular leading edge of a clock. For example, a micro-controller might have to have several control lines ready at precisely the leading edge of the number one phase of clock micro_clk. This could be written as:

```
r := microstore [micro_ip];
micro_ip := micro_ip + 1;
ASSIGN r.carry TO line1 SYNC micro_clk PHASE 1;
ASSIGN r.shift0 TO line2 SYNC micro_clk PHASE 1;
ASSIGN r.shift1 TO line3 SYNC micro_clk PHASE 1;
ASSIGN r.shift2 TO line4 SYNC micro_clk PHASE 1;
ASSIGN r.clear TO line5 SYNC micro_clk PHASE 1;
(*etc.*)
```

All of the above net updates will be effected at precisely the same simulated time.

The user may specify any number of independent clocks, each with its own period and number of phases. Unlike some other simulation environments, clocks in ADLIB do not consume any computation resources themselves; only if and when a component accesses them is any calculation performed. The user may mark one of the clocks as being the default. This saves him or her from writing the clock's name in every SYNC clause. If the keyword PHASE is not specified, phase 0 is assumed. These facilities make it convenient to describe systems that maintain a single clock, such as a pipelined multiplier that keeps each stage in lock step with the others.

The default net update mode schedules one new event and performs no operations on other events. This mode is useful for high-level modeling where many of the timing details associated with gate-level designs do not exist. In these situations, simplicity of modeling and efficiency of event-queue processing is desirable, and the default mode is the most appropriate.

Another assignment mode, called *transport*, can be specified to override the default mode. In transport mode a new net update event is scheduled, and then all events associated with the same net that are set to occur after the specified update time are removed. This mode is important for handling gate-level devices which have more than one propagation delay. Without it, it is possible to get an incorrect output. This can occur when the difference between a rising and falling propagation delay results in several spurious updates being scheduled for the output of a gate, and the last output scheduled does not reflect the correct final value. The transport mode prevents this by suppressing all future events on each update.

2. The WAITFOR statement,

    WAITFOR <boolean expr> <control clause>

The WAITFOR statement causes the execution of a process to pause, and does not allow it to continue until <boolean expr> evaluates to true. The <control clause> may come in one of two forms. First, a <timing clause> may be used, just like the timing clause in an ASSIGN statement. If a delay clause is used the <boolean expr> is reevaluated periodically at the period specified in the delay clause. For example:

    WAITFOR current>0.001 DELAY sample_period;

This statement checks the value of current every sample_period time units, until it exceeds one milli-amp.

If the SYNC keyword is used, then <boolean expr> is reevaluated each time the specified clock goes through the specified phase. For example:

```
WAITFOR acknowledge=1 SYNC bus_clock PHASE 4;
```

This statement would not allow execution to continue until the net acknowledge was equal to 1 on the leading edge of the fourth phase of clock bus_clock.

Alternatively, a control clause may take the form of a list of ports to which the component is to be sensitized. This format is called a *check list* because whenever any of the ports mentioned in it is updated, the boolean expression is rechecked, as in:

```
WAITFOR data_rdy = 1 CHECK data_rdy;
```

This statement would put the component into a passive state until the net data_rdy was updated to the value 1. No simulation resources are consumed while the component is idle; there is no busy waiting. In particular, if other components try to update data_rdy with its current value, the boolean expression is not reevaluated. This is because the simulation environment automatically deletes all such null updates.

3. **Sensitize, Desensitize, and DETACH**

Taken collectively, these provide a facility for direct control of the operation of a component. They operate in a way that is similar to WAITFOR, but at a lower level and somewhat more efficiently. Sensitize and desensitize are predefined procedures which make a component receptive or immune to changes on its inward nets. These procedures are always used with the DETACH operator, which causes execution of a component to stop until one or more of the nets to which it is sensitive is updated. When an update on a sensitized net occurs, the component will be awakened. These three statements may not appear in subprocesses.

Because of the flexibility of the WAITFOR construct, it is difficult to think of an application where DETACH is really more convenient and not merely more efficient. One possible exception is finite-state machines. Here DETACH is used to describe a finite-state machine that recognizes the bit strings consisting of ones and zeros. The strings must match a regular expression that begins and ends with 1, and where any 0 must be preceded and followed by at least one 1 (example taken from [KOH70]). This machine is stimulated by a net called input_line, which contains a data element d and a strobe field s. (To drive this machine, it is necessary to put the data value in the d field, and to update the s field.) In ADLIB, one way to define the automaton is:

```
sensitize (input_line);
1: DETACH; (*initial state*)
       IF input_line.d = 0 THEN GOTO 3;
2: (*accepting state*)
       writeln(output,'accept');
       DETACH;
       IF input_line.d = 0 THEN GOTO 1;
       GOTO 2;
3: DETACH; (*terminal state*)
       GOTO 3;
```

4. The wait and signal procedures.

    The syntax for these procedures are:

    wait(<semaphore>)

and

    signal(<semaphore>)

Semaphores are used to synchronize concurrent processes, and to force key sections of code to be executed before others. When the wait function is encountered, the process will go to sleep until another function encounters the signal function with the same semaphore argument. An example is given later in the discussion of semaphore net types.

5. The UPON subprocess.

    The syntax for an UPON subprocess is:

    UPON <boolean expr> <control clause> DO <stmt>;

UPON is used to define a set of activities to be performed independently of the main activities of the component. If <control clause> contains a list of nets then whenever one or more of those nets is updated <boolean expr> is reevaluated. If it is true, then <stmt> is executed. For example:

```
interrupt: UPON (interrupt.priority > current)
  CHECK interpt DO
      BEGIN
      push(machine_state);
      service_interrupt;
      pop(machine_state);
      END;
```

This code checks the priority level whenever the net interpt was updated, and services the interrupt when necessary.

6. The TRANSMIT subprocess.

The syntax is:

```
TRANSMIT <expr> <control clause>
     TO <port name> <timing clause>
```

TRANSMIT is more specialized than UPON. If the control clause contains a list of ports, then whenever one or more of the nets attached to these ports is updated, <expr> gets reevaluated. The result is then assigned to <port name> at the time specified by <timing clause>. TRANSMIT is very convenient for describing combinational circuitry. For example, a simple NAND gate can be described with:

```
nand:TRANSMIT NOT(a AND b) CHECK a,b TO c DELAY 15;
```

7. The inhibit and permit procedures.

The name given to a subprocess may be used to control it by means of the procedures permit and inhibit. All subprocesses are initially un-inhibited, which means that any external stimulus can activate them. A component may inhibit some or all of its subprocesses, thereby making them insensitive to stimuli. Subprocesses may be permitted or inhibited at any time. For example, a computer may protect a critical region with:

```
inhibit(interrupt);
write_to(shared_data);
permit(interrupt);
```

A subprocess that is inhibited has no memory of its previous state; when it is permitted to run again, its control clause must be completely reevaluated. For example, if DELAY 5 were specified, then the subprocess will be activated 5 time units after it is permitted, regardless of the time when it was inhibited.

3.4 Designing an ADLIB Program

To illustrate how ADLIB is used, this section describes the process used to design a small system that plays Blackjack. This was inspired by a DDL design found in [DIE75], but is somewhat more complex and complete. The ADLIB system consists of one dealer and one or more players. Nets represent the flow of cards from the dealer to the players, and coordinate player activities. To begin the design process, one must first consider what data needs to be transmitted (suit and rank), and what control information is must pass between the player and the dealer, *i.e.*. player is waiting for card, dealer is waiting for player, *etc.* This information can be encapsulated in three types of nets, as shown here:

```
PROGRAM cardgame;
     TYPE
     suit_type = (clubs,hearts,diamonds,spades);
     rank_type = (ace,two,three,four,five,six,seven,
          eight,nine,ten,jack,queen,king);
     NETTYPE
     card_bus = RECORD
          suit : suit_type;
          rank : rank_type;
          END;
     display_lights = (hit,stand,broke);
     control_line = (card_rdy,card_accepted);
     COMPTYPE dealer; (* deals out cards*)
          BEGIN
          (* not yet designed*)
          END;
     COMPTYPE player; (*accepts cards, stands or goes broke*)
          BEGIN
          (* not yet designed*)
          END;
     BEGIN
     END.
```

Figure 3-2. Outline of Blackjack System

The net types shown fall into two categories: structured net types, such as card_bus; and scalar net types, such as display_lights and control_line. Structured net types are most useful when several pieces of information are logically associated but need to be updated and examined independently.

```
(* player's code *)
WAITFOR cntrl = card_rdy CHECK cntrl;
(* accept card*)
ASSIGN card_accepted TO cntrl;
(* process card, go broke, hit or stand*)

(* dealer's code *)
(* generate next card *)
(* ASSIGN nextcard to card_bus *)
ASSIGN card_rdy TO;
WAITFOR cntrl = card_accepted CHECK cntrl;
```
Figure 3-3. Code Fragments for Coordination

The code in Figure 3-2 specifies three net types but no control protocol. One control line runs between each player and the dealer that serves it. Its value alternates between card_rdy and card_accepted. The synchronization mechanism is expressed by the code fragments shown in Figure 3-3. The first WAITFOR statement shown causes the player to wait until the card is ready, and the second causes the dealer to wait until the player has decided what to do.

The design is now ready for the specific algorithms used by the players and dealers.

```
COMPTYPE player;
        (* declare the ports of this component*)
INWARD
        card : card_bus;
OUTWARD
        lights : display_lights;
EXTERNAL
        cntrl : control_line;
        (*declare the storage needed by this component*)
VAR
        score : 0..27;
        holding_ace : boolean;
BEGIN
WHILE true DO
        BEGIN
        holding_ace := false;
        score := 0;
        REPEAT
                REPEAT
                        ASSIGN hit TO lights;
                        ASSIGN card_accepted TO cntrl;
                        WAITFOR cntrl=card_rdy CHECK cntrl;
                        IF card.rank < jack THEN
                                score := score + ord(card.rank) + 1
                                (*ord returns an integer in 0..13*)
                        ELSE
                                score := score + 10;
                        IF (card.rank=ace) AND (NOT holding_ace)
                        THEN
                                BEGIN
                                score := score + 10;
                                holding_ace:=true
                                END;
                        UNTIL score >= upper_limit;
                IF (score > 21) AND holding_ace THEN
                        BEGIN
                        score := score - 10;
                        holding_ace := false
                        END;
                UNTIL score >= upper_limit;
                If score <= 21 THEN
                        ASSIGN stand TO lights
                ELSE
                        ASSIGN broke TO lights;
                END;
        END;
```

Figure 3-4. Definition of Component Type Player

For the purpose of this discussion the player accepts cards until it reaches its limit, treating aces as 1 or 11 points as needed. The ADLIB code for the player is shown in Figure 3-4.

In order to see if component type **player** works properly it is necessary to develop a **dealer** component type to drive it. There are several possible ways to do this, just as there are several ways to test a new piece of hardware. In ADLIB, it is easy to write a component type that interacts with the terminal. Another possibility is to use a pseudo-random number generator to choose cards. A collection of such generator subroutines called **rndpak** is available to ADLIB users. Finally, the designer can write a component type that reads the test data from a file. Each of these approaches is valuable at different phases in the design process. One approach might be to use interactive testing for initial debugging, large numbers of random inputs for extensive testing, and selected prerecorded values for production.

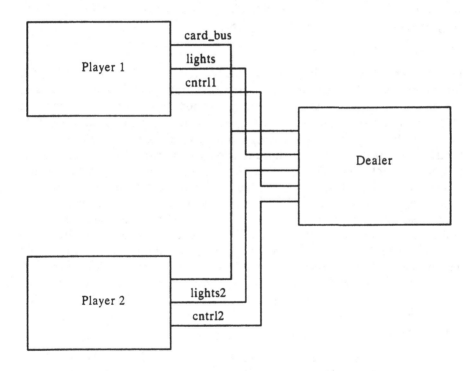

Figure 3-5. Structure of a Simple Cardgame

Now that the behavior of component type **player** and **dealer** have been specified, any number of players and dealers may be used in the structural part of the design. The structure of one possible cardgame is shown in Figure 3-5.

3.5 Types, Net Types and Type Checking

Like Pascal, ADLIB is a *strongly typed* language [OEI78]. This means that each piece of data, each function and all parameters must be declared with exactly one type. Each use of a variable or net must be *type-compatible* with its declaration. In a well-written program the type of a variable defines the narrowest possible range of values that it may attain. Typing is essentially a way for the designer to express his or her intentions about the way in which a piece of data should be used. The compiler can automatically detect when those intentions are violated, which usually implies an error. In languages that are not strongly typed it is often easy to treat characters as integers or treat integers as pointers. Even when these uses are intentional and operate correctly, the source code is difficult to read, understand and maintain. And when it is done unintentionally, chaos can result. A discussion of Pascal typing and program reliability is provided in [WIR75].

3.6 Type Checking of Nets

The primary type checking mechanism in ADLIB is the use of net types for defining the intended interconnection mechanism between components. In the structure specification the nets that connect components must be declared with identical net types at each end. This net type checking is more thorough than checks found in most register transfer languages (RTL's), where it is only necessary that the number of bits must match. To illustrate this difference, consider a component that produces two BCD (binary coded decimal) digits and another that accepts eight bits of binary data. Most RTL's would allow them to be directly connected since both are eight bits wide. Even simulation might not detect this potential error if the test set did not happen to include any values greater than 9. But the ADLIB simulation environment would detect the mismatch, since type BCD is not type compatible with type binary. If the designer actually intended the interfaces to be compatible, a *translator* component must be provided. Translators provide the ability to do multi-level simulation at the expense of some type-checking security and some loss of data precision.

Although they can be used to catch errors, types should not be viewed just as a restriction. Because the set of types can be extended by the user arbitrarily, they provide a mechanism to tailor the language to the problem, and can actually simplify the design process. The next section illustrates some ways that a designer can take advantage of net types to reduce errors and improve readability.

3.7 Data Types Available in ADLIB

Because ADLIB is a superset of Pascal, it inherits all of the Pascal type construction mechanisms. These types come in two basic categories: scalar and structured. Scalar types carry a single value, structured types can carry many values. Some types are predefined, additional ones may be constructed by the user. For the benefit of readers not familiar with Pascal's types and type constructing mechanism, they are summarized here:

1. Integer

   Type **integer** is a predefined scalar type with a range that is machine dependent, but which includes the most negative and most positive integer that can be represented.

2. Character

   Type **char** is the predefined scalar type for describing character data.

3. Real

   Type **real** is the predefined scalar type used to describe floating point numbers.

4. Enumerated Types

   Pascal and ADLIB allow a user to enumerate (list out) all the possible values of a piece of data. Such an enumeration produces an enumerated type. For example:

   ```
   logic_level=(low, high, unknown, high_impedance);
   ```

   Several examples of enumerated types were used in the cardgame example earlier in this chapter. Enumerated types are also useful for describing the instruction sets of machines such as the PDP8:

   ```
   TYPE
     opcodettype=(xand,tad,isz,dca,jms,jmp,iot,encoded);
     mops=(cla,cll,cma,cml,rar,ral,rt,iac,sma,sza,
     snl,is,osr,hlt);
   ```

   In ADLIB, one can use a CASE statement to describe the execution of a machine instruction in a format similar to ISP [BCG71]. For example part of the ADLIB code to describe the PDP-8 is shown below:

```
CASE opcode OF
     (*routines that start with rg are part of rgpack*)
     xand: acc := rgand(acc,m[iz]);
     tad: BEGIN
          acc:= rgadd(acc,m[iz]);
          IF rgcarry <> link THEN
                  link := 1
          ELSE
                  link :=0;
          END;
     isz: BEGIN
          (*simulate extra memory cycle*)
          WAITFOR true DELAY 1;
          incr(m[iz]);
          IF m[iz]=0 THEN incr(pc);
          END;
     dca: BEGIN
          m[iz]:= acc;
          acc := 0
          END;
     jms: BEGIN
          m[iz]:= pc;
          pc := iz+1
          END;
     jmp: pc:= z;
     iot: IF trace THEN
                  dump(iot,[dcacc])
          ELSE
                  BEGIN timer:=timer+1;
                  IF timer>200 THEN stopsim;
                  END;
     encoded: BEGIN (*etc*)
```

5. Boolean

Type boolean is a predeclared enumerated type useful for
describing logical states. Its definition is

```
boolean = (false,true);
```

For example, if strobe were declared to be of type boolean, one could write:

```
strobe := data_rdy AND (bus_clock = 3)
```

6. Subranges

Pascal and ADLIB allow the user to declare types that are subranges of scalar types declared previously. These specify that only part of a range of scalar values is acceptable, for example:

```
register_number = 0..7
```

specifies not only that variables of type register_number are integers, but also that they must lie between 0 and 7 inclusive. Assigning a value to a subrange variable that is outside its range is automatically detected. For example, in the ADLIB blackjack machine the variable score was declared to range over the values 0 to 27, the high value being equal to 16 (the highest possible score before standing) + 11 (value of an ace). By contrast, the DDL version merely declared the score to be a five bit register. The ADLIB approach has two advantages. First, it allows the designer to defer any decision on the representation of data in the early stages of design. Second, it encodes more information: *e.g.* the fact that the score can never exceed 27. Although this is not too critical here, it is easy to visualize applications where knowledge of the range of data that a register holds can be used to improve the design. For example, it might be useful to know not only that memory addresses in a DEC-10 are 18 bits long, but also that they range from 16 to 262144 since the first 16 addresses refer to registers.

7. Arrays

Arrays are structured types used to describe homogeneous sets of variables, (similar to DIMENSION statements in FORTRAN). For example:

```
memory = ARRAY[0..1023] OF integer;
```

Since Pascal and ADLIB do not have a string data type, the only way to manipulate strings is with arrays of characters.

8. Records

Records are structured types that are useful for grouping related but not necessarily homogeneous variables, as for example:

```
printerline = RECORD
        length : 0..132;
        text = PACKED ARRAY [1..132] OF CHAR;
        END; (*of record*)
```

Most register transfer languages allow the user to define several names that refer to the same piece of data. One example of the utility of this is in describing an instruction register, where one of the bits is given a mnemonic name such as "I" (for Indirect) in addition to being "IR[0]." This can make parts of a program more readable, but can also lead to confusion when a mnemonic is referenced for the first time several pages away from its declaration. The strategy adopted by ADLIB is to use variant records for this purpose. A variant record is a single data area that may have several different data structure templates applied to it. One example of this is the four ways that a programmer can look at an HP 2116 instruction, as discussed in the 2116 machine manual [HP]. In ADLIB, these alternative views would be encoded as:

```
TYPE
    instr_variant = (whole,memory_ref,
        register_ref,i_o);
VAR
    ir : RECORD
        CASE instr_variant OF
            whole : (ARRAY[0..15] OF bit);
            memory_ref : ( indirect : bit;
                mem_instr : ARRAY[0..4] OF bit;
                zero : bit;
                mem_addr : ARRAY[0..9] OF bit);
            reg_ref : (group : ARRAY[0..3] OF bit;
                micro : ARRAY[0..11] OF bit);
            i_o : (io_group : ARRAY[0..3] OF bit;
                io_instr : ARRAY[0..5] OF bit;
                select : ARRAY[0..5] OF bit);
        END; (*record*)
```

This record informs the reader and the compiler that the instruction register ir may be viewed in four different ways, but is still really just one register 16 bits long. (Note that the total number of bits in each variant is 16.) Access to ir can then be performed using the mnemonic fields, as for example:

```
ir.whole := data_bus;
```

or

```
IF ir.indirect = 1 THEN cycle := fetch;
```

Now the fields are closely associated with the register and are less likely to be misinterpreted.

9. Sets

A SET variable is a structured variable that may contain from 0 to all of its members, *i.e.* a powerset. Sets are delimited textually by "[" and "]", and facilities are provided for set intersection (AND), union (OR), difference (-) membership (IN), equality (=), and size comparison (< and >). Since sets are often packed into machine words these operations are usually efficient. Sets are convenient for visually grouping related symbols for the reader and may improve the efficiency of simulation. For example, in the 8008 we can express certain facts in machine readable form that are normally only shown on the data sheets, such as the grouping of instructions into categories:

```
index_instructs:=[lrr,lrm,lmr,lri,lmi,inr,dcr];
 one_cycle_alu:=[adr,acr,sur,srr,ndr,xrr,orr,cpr];
```

As an example of the use of set operators, consider the code that describes the timing of part of the execution cycle. It might contain:

```
IF instruction IN one_cycle_alu THEN
    WAITFOR SYNC PHASE 1;
```

10. Files

The most familiar examples of file variables are the predeclared files input and output. These files are of type char, which is the most common type of file. However, the Pascal/ADLIB user may declare additional types of files to match the data to be stored in them. Storing and retrieving that data can then be accomplished efficiently. For example, it is easy to describe a core image as:

```
core_image : FILE OF integer;
```

For convenience, special facilities are provided for reading and writing files of text.

11. Pointers

Pascal and ADLIB provide two independent areas for storing data: the ordinary stack and a heap. The heap is accessed only via special pointer variables, which may in turn point to other pointers, *etc.* This makes it convenient and efficient to develop complex data structures. Pointers are denoted by the up arrow "^". Normally, such complex data structures are used in component definitions describing software. For example, pointers are essential for describing linked lists. One way to define a link is by combining the record and pointer facilities, *i.e.* :

```
link = RECORD
      prev, next : ^link;
      data : integer;
            (*any data structure may go here*)
      END;
```

If `linka` and `linkb` are pointers to links, then they can be connected to each other

```
linka^.next := linkb;
linkb^.prev := linka;
```

New elements can be added to the heap by using the predeclared procedure new. For example, to make `linka` point to a new link one would write:

```
new(linka);
```

Normally, complex data structures using pointers are used to model software, not hardware, since there is little direct correlation between them and physical devices. However, it is important that such facilities be available if ADLIB is to be used at all levels of design. In addition, they add flexibility to the language that makes it possible to write ADLIB models that support more than just functional or logic simulation, as will be discussed in Chapter 5.

Whereas the previous types are part of both Pascal and ADLIB, the following two are available only in ADLIB.

12. **Bit**

This is a predeclared subrange of integer. It may range over the values 0..1.

13. **REGISTER**

The **REGISTER** net type is used to model register level circuits. A register can be thought of as a string of bits, with a virtually unlimited length (as many as 2000 bits in some implementations). For example, a 64 bit machine could be simulated on a 32 bit computer via the use of registers. In addition, many of the operations normally available at the register level in circuit designs are available in the ADLIB language, simplifying the task of writing a model.

The length and bit numbering conventions for a given register are
specified when the register is declared. For example, the
declarations for several register variables are shown here:

```
VAR
        rega: REGISTER [1..5];
        regb: REGISTER [11..0];
        regc,regd: REGISTER [21..13];
        rege: REGISTER [-1..1];
```

Figure 3-6. Register Declaration

In this case, a register variable rega is declared to have a length of
5 bits, with an indexing scheme starting at 1 and increasing to 5. The
right-most bit is the Most-Significant Bit (MSB) with an index of 5,
and the Least-Significant Bit (LSB) has an index of 1. Similarly, the
variable regb is declared to have 12 bits, with the MSB numbered
0, and the LSB numbered 11. Finally, the regc variable is declared
to have 9 bits with the MSB numbered 13 and the LSB numbered 21.
Indexes can have negative values as well as positive values, as
shown for register rege.

```
regc:=regd;
regb[11..9]:=rega[1..3];
rega[1..5]:=regc[21..20]//regd[21..19];
rega[1..1]:=regd[13..13];
```

Figure 3-7. Assignment of Register Variables

Figure 3-7 shows several examples of register value assignments.
Each variable shown above corresponds to the declarations of
Figure 3-6. In the first statement, the entire contents of regc are
copied to regd. In the second statement, three bits of regb are
given the values associated with three bits of rega. In the third
statement, two strings of bits associated with registers regc and
regd are concatenated and assigned to a portion of rega. In the
final statement, a single bit is assigned into rega.

```
rega
contents          1   1   0   0   1
                  ---------------
indices           1   2   3   4   5

rege
contents            1   1   1
                    ---------
indices            -1   0   1
```

Figure 3-8. Example Register Values

```
        rega:=rege;

rega
contents          0   0   1   1   1
                  ---------------
indices           1   2   3   4   5

        rege:=rega;

rege
contents        0   0   1
                ----------
indices        -1   0   1
```

Figure 3-9. Register Assignment

If the receiving register is not large enough to hold all the bits, the left-most (least significant) bits are truncated. If the receiving register can hold more bits than are assigned, zeros are put into the right-most (most significant) bit positions. Referring back to the declarations of Figure 3-6, assume the contents shown in Figure 3-8. Several assignments and corresponding results are given in Figure 3-9 given the starting values shown in Figure 3-8.

14. Semaphore Net Type

The **semaphore** net type controls the interaction between concurrent processes. Its definition is shown here:

```
NETTYPE
        semaphorenet:SEMAPHORE;



...
INWARD a,b:boolnet;
INTERNAL phase_2_start:semaphorenet;
SUBPROCESS
        proc1:  UPON TRUE CHECK a DO
                BEGIN
                (* ... phase 1 ... *)
                WAIT(phase_2_start);
                (* ... phase 2 ... *)
                END;
        proc2: UPON TRUE CHECK b DO
                BEGIN
                SIGNAL(phase_2_start);
                (* ... *)
                END;
...
```

Figure 3-10. Use of Semaphores

Semaphores are used with the WAIT and SIGNAL operations, as shown in Figure 3-10. In this figure, the internal port phase_2_start has been used to synchronize the execution of the phase 2 code with the activation of the proc2 process.

3.8 Another Example: The Rs232 Interface

One can combine enumerated types with records to create a strong and specific definition of an interface. For example, consider the rs232 connection standard. Most RTL's would merely specify it as a 25 bit connection, which could be written in ADLIB as:

```
rs232 = ARRAY [1..25] OF bit;
```

However, this would not make best use of the facilities available. In ADLIB, it is possible to write:

```
TYPE
widerange = (neg_12V,pos_12V);
grounded = (zero);
NETTYPE
      rs232 = RECORD
              fg : grounded; (*frame ground*)
              td : widerange; (*transmit data*)
              rd : widerange; (*received data*)
              rts : widerange; (*request to send*)
              (*etc.*)
              END;
```

If the net tty1_line were declared to be of net type rs232, then the compiler would accept

```
ASSIGN neg_12V TO tty1_line.td;
```

but would flag as an error:

```
ASSIGN 0 to tty1_line.terminal_rdy;
```

because of type incompatibility.

3.9 Contour Models

The *scope* of an identifier is the range over which that identifier is visible. A *contour model* is a two-dimensional graphical representation of scopes; it uses one rectangle to represent each scope. etc. Such a drawing can be used to answer questions about identifier visibility, naming conflicts, and data hiding. (For readers unfamiliar with contour models, an introductory tutorial is available in [JOH71]. Also of interest is *SIMULA Begin* [BIR73] which is tuned to SIMULA67, and OREGANO [BER71]. However, the basic principles of contour models are simple, and this description will avoid the more complex issues.) The set of identifiers visible at any point in the model is determined by examining each enclosing contour in turn. For example, in the model that describes a PACKAGE, the outermost contour represents the predeclared identifiers, such as integer and time, the next one in is the user's global identifiers. Inside that are subroutines, which may be recursive. Thus if an identifier appears inside a subroutine, the search for its meaning begins in that routine, then goes on to the user defined globals, and finally the predeclared identifiers. Identifiers inside non-nested contours can be accessed only through *access pointers* that link one scope with another. Access pointers are used extensively in other languages, such as SIMULA67, to allow one process to reach into another and alter internal attributes of it. But this facility is not available to the ADLIB user directly, because it invites hard-to-detect side effects and bad code structuring.

3.10 Packages and Modules

The ADLIB language, as supported by HELIX, allows the modular development of model libraries and subroutine libraries through the use of the *package file* and the *module file*.

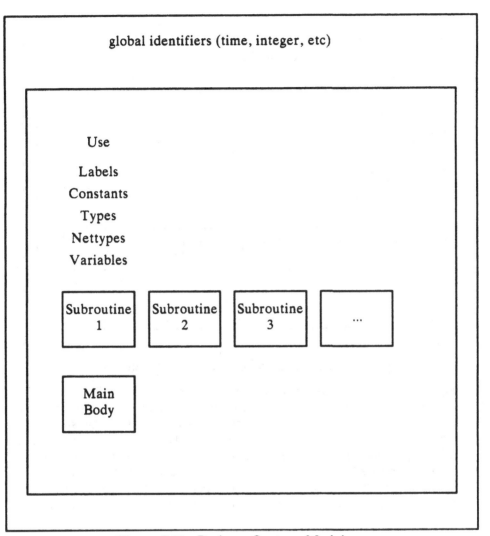

Figure 3-11. Package Contour Model

Figure 3-11 shows the overall structure of a package file. A package file comprises a self-contained set of procedures and an associated global initialization body. These procedures can be referenced by any ADLIB model via a USE statement.

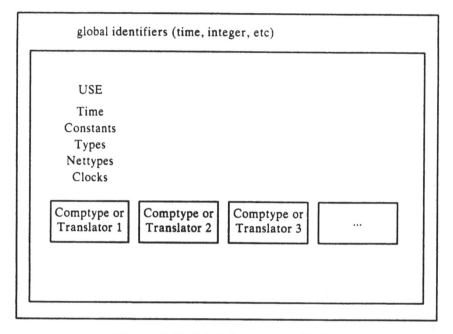

Figure 3-12. Module Contour Model

Figure 3-12 shows the structure of the module file. A module file contains a set of related component types and translators. The purpose of a module file is to allow related collections of models to be stored in a single file. In addition, models can be split between several module files, allowing greater flexibility in compilation and development of models. Through the USE statement, any module file can include all subroutines associated with a package file. Since packages are precompiled, this approach is efficient, and allows modular development of subroutines.

3.11 Global Identifiers

The models shown in Figure 3-11 and 3-12 are not much different from the contour model of a Pascal program. In the upper left hand corner of each rectangle appear the user-defined labels, constants, and types and also the user-declared variables*. In addition, more rectangles may

---

* In this book, items that do not consume storage at runtime are said to be *defined*, and items that do are said to be *declared*. For example, subroutines are defined, variables are declared. In particular, items that are defined do not appear inside contours in the dynamic models.

appear within a rectangle representing nested scopes.

In the outermost contour is the predeclared variable time which represents the simulation time. User assignment to this variable is illegal and is detected during compilation. Contour 2 in Figure 3-12 is the most global level for the user. In it are found the user's global labels, constants, types, net types, clocks, variables, subroutines**, and component types. The meanings of labels, constants, and type definitions are unchanged from Pascal. There is also an algorithm associated with this contour that could be called the *main body* of the program; it can be used for initialization of global variables, resetting files and such. This main body may not contain any ADLIB control primitives such as DETACH or WAITFOR, and may not access any component or net. During the execution of the program's main body, time and all clocks are have the value zero. This code block may call global subroutines that call further routines recursively, just like the main body of an ordinary Pascal program.

The inclusion of global variables into ADLIB is a concession to practicality and user convenience. Ideally, a design should not have any, since they might represent inter-component connections that have no physical correspondence. In practice, however, such variables tend to be useful for several important functions, including data reduction (such as statistics gathering), generating simulation input (with a common input file), and even for semantically interesting things like semaphores. (The last is possible because an access to a global variable appears to be an indivisible operation to the rest of the simulation). On the other hand, global variables should not be used for intercomponent communication. This is what ports are intended for, just as parameters are intended for communicating with subroutines. It may be possible to detect such clandestine component interaction during compilation and prohibit it, just as there have been proposals to ensure that subroutines have no side effects. However, such mechanisms can generally be defeated, and are invariably unpopular with programmers. The decision of how to use global variables is therefore left to the user.

3.12 Net types

Net type definitions are similar to type definitions, except that they are matched to port declarations, instead of variable declarations. Likewise,

---

** Throughout this book, the word "subroutine" or "routine" is used to mean procedure or function.

port declarations are similar, but not identical to variable declarations. Ports are allocated in a different way from variables, and are interconnected with other components and with the simulation support system. The various net types thus define the ways that the components are able to interact. During simulation, the support system generates new data items of the various net types, compares them, does assignments to them, and dynamically regenerates the storage allotted to them. Net types (as opposed to ordinary types) must be used whenever ports are declared on component types or are used as parameters to subroutines. Functions may not return ports. To update a port, the ASSIGN statement and the TRANSMIT subprocess facilities are provided*. (This is somewhat like SIMULA67, where updating a pointer requires a special syntax.) Any expression assigned to or compared with a port must be type compatible with the net type of the port. Within a component type, two types or net types are compatible if they are subranges of the same base type. This is not true in connections between components, because each net type is incompatible with all the other net types. (ADLIB is a type-by-name language.) This strengthens the inter-component type checking because it allows the designer to distinguish net types that are subranges of the same base type.

3.13 Clocks

Following the net type definitions, an ADLIB program may contain one or more clock definitions. A clock function may be invoked anywhere that a non-constant integer expression is allowed.

---

* In the original ADLIB, ports may not appear on the left side of assignment (":=") statements. In HHDL this is allowed, because it has several practical uses. However, it should be done only with a thorough understanding of the simulation environment, because it can lead to hard-to-trace errors.

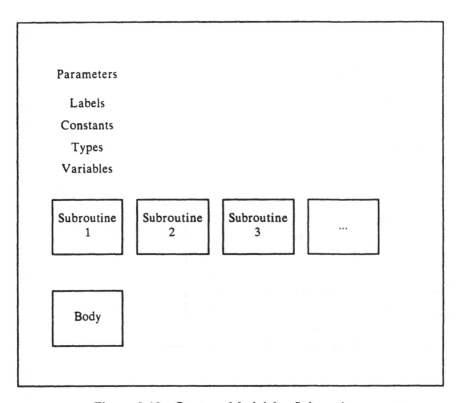

Figure 3-13. Contour Model for Subroutines

3.14 Subroutines

Figure 3-13 shows the structure of subroutines. Global subroutines are represented as in Pascal and may include more subroutines nested within them. These subroutines may be freely called from inside any component type definition, and may contain net update statements to port parameters, which may be visible directly or as parameters. However, routines may not contain WAITFOR or DETACH statements. This restriction allows an enormous simplification and acceleration of the runtime support because it sharply reduces the need for dynamic storage reclamation ("garbage collection") found in some simulation languages, (*e.g.* SIMULA67). ADLIB provides other facilities such as subprocesses and interprocess communication which compensate for this.

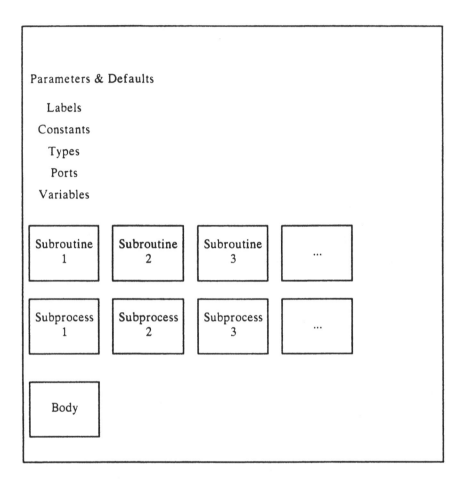

Figure 3-14. Component Type Definition Contour Model

3.15 Component Type Definitions

Following the global subroutine definition section is the central part of the ADLIB program, the component type definitions. The other constructs, such as the clock definitions, serve only to support them. Component type definitions are similar to subroutine definitions in that they define algorithms for data manipulation. The structure of a component type definition is outlined in Figure 3-14. It includes parameters, defaults, port declarations, labels, constants, types, variables, subroutines, subprocesses and the main body of the component type.

3.16 The Heading of a Component Type Definition

A component type definition's parameters are similar to those of a routine, except that they may be set only in the structure definition language. For example, a component type **nandgate** might have a parameter **risetime**. This would enable many instances of the nand gate to be allocated by the simulation environment, with each one potentially having a different rise time. This is simpler and more efficient than requiring separate models for each. The default section, which follows immediately after the parameter list, may be used to specify default values for any or all of the parameters. The parameters listed in the default section do not have to be in the same order as in the parameter list. All parameters to component types are call-by-value. Pointers, files, and structured data types are prohibited.

3.17 Port Declarations in a Component Type Definition

Following the parameter default section is the declaration of the ports used by the component type. These act as the interface between the component and its environment. Ports must be marked as one of the following: INWARD (receive data only), OUTWARD (transmit data only), and BOTHWAYS (both receive and transmit). In addition, the user may declare INTERNAL signals which can be assigned and evaluated, but only from within the component that declares them. It is illegal to ASSIGN or TRANSMIT to an INWARD port, to sensitize an OUTWARD port, or to place an OUTWARD port after the keyword CHECK. The intention here is to ensure that information never flows from a port marked OUTWARD or to a port marked INWARD. If both forms of access are needed the port should be marked BOTHWAYS.

3.18 Internal Signals

INTERNAL signals are provided in ADLIB to facilitate the description of components with complex behavior or timing characteristics. These provide a mechanism for components and subprocesses to wake themselves up, store and retrieve information and accurately model inertial delays. Although internal signals must be declared with a net type, they are not ports, but rather are part of the behavior specification only. They do not appear in any structural description because they have no correspondence to any physical data path. They are more like local variables to the component, and can be used for holding temporary values.

For example, when specifying an instruction set it might be necessary to define a swap instruction that exchanges two registers a and b. If a and b were ordinary variables, then the statements:

```
a := b;
b := a;
```

will result in both being set to the original value of b, which is an error. A temporary value is needed to perform the swap correctly. One way is to define a pseudo-register called t and write:

```
t := b;
b := a;
a := t;
```

However, this has several disadvantages. First, it introduces another register that is not really part of the component being described (the reader may think it is available for general use). Secondly, it overspecifies the order in which the transfer is to take place. (In fact, both registers may be updated simultaneously). Finally, it does not express any timing for this operation. All these problems can be solved with the use of internal signals. If a and b are internal signals the operation can be described as:

```
ASSIGN a TO b SYNC;
ASSIGN b TO a SYNC;
```

The simulation environment environment will automatically allocate temporary storage so that this operation will be performed correctly. The updates will take place simultaneously with phase 0 of the default clock.

3.19 Standard Pascal Declarations

Labels, constants, types, variables and subroutines in component types are unchanged from Pascal, and again these subroutines may be nested arbitrarily. Following the normal scoping rules, routines have access to all the identifiers inside the component type, and to all those defined at the global level.

3.20 Subprocesses

The next part of a component type definition is the subprocess declaration section. Subprocesses are like little components that run autonomously from the main component body, but under its control. They can be used, for example, to describe the direct memory access channels in a mainframe computer. Their purpose is to simplify the code in the main body of the component type by taking care of secondary functions.

3.21 The Main Body of a Component Type Definition

The main body of the component type describes the fundamental activities of that type of component. For simple components, it may contain all the behavioral information directly. For more complex components, some designers find it more convenent to break up the behavior into the subprocesses, and use the main component body only to initialize and control them. The main body itself has the added flexibility over the subprocesses that it may also place itself into a wait state, where it stays until some stimulus is received or some condition is met.

The contour model of a component type shown in Figure 3-14 shows subprocesses (UPON and TRANSMIT statements) represented by their own contours within the component type. These hold the algorithm of the subprocesses but no variables. The main body of the component type is also enclosed within its own contour.

3.22 Static Structure of an ADLIB Program

The static contour model for an ADLIB program in the original DEC-10 implementation is illustrated below.

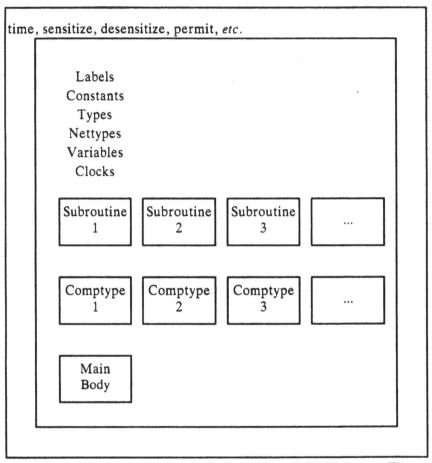

Figure 3-15. Static Contour Model of an ADLIB Source File

3.23 Summary – How Languages are Defined

The definitions of most languages, computer or human, depend on a combination of formal specifications for some aspects, such as syntax, complemented with less formal prose descriptions and examples. This chapter has informally sketched out some aspects of ADLIB, including its basic statements and data types, its rules for scoping, and some simulation semantics. For most readers, this should be sufficient for them to follow the examples in the next two chapters. For readers concerned with the details of syntax, a formal BNF is provided in the Appendix. Finally, a somewhat more rigourous definition of the dynamic semantics is available in the original ADLIB thesis [HIL80].

CHAPTER 4

EXAMPLES OF DESIGNING WITH ADLIB

This chapter describes several electronic design examples, and illustrates how the ADLIB language is used for each. A simple gate-level circuit is the first example, and is followed by more complex problems that illustrate the expressive power of the language and demonstrate how it can be applied in a variety of situations.

4.1 A Gate-Level Example

As previously mentioned, both a structural and a behavioral description is required to simulate a circuit using the ADLIB language. Such a structural description is given in Figure 4-1, which is a schematic diagram for a clocked RS flip-flop. This schematic diagram describes the connections among four NAND gates, and from this circuit to the outside world, but does not specify the detailed behavioral characteristics of the gates themselves. Such information must be specified independently, perhaps through part numbers that index into a manufacturer's data book. Only then would the behavior of this circuit be fully specified.

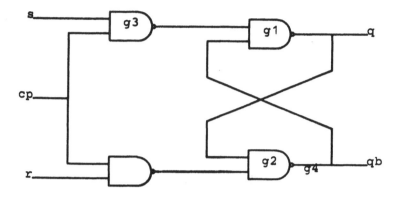

Figure 4-1. Clocked RS Flip-Flop

```
(* SDL Description for NAND Symbol *)
NAME: nand;
(* Define External Pins on Symbol *)
EXT:: a,b,y;
INPUTS: .a,.b;
OUTPUTS: .y;
END;
ENDC;

(* SDL Description for RSFF Circuit *)
NAME: rsff;
(* Define External Pins for Circuit *)
EXT:: cp,s,r,q,qb;
INPUTS: cp,s,r;
OUTPUTS: q,qb;

(* Declare Referenced Symbols, and Instance Names *)
TYPES: nand;
nand: g1,g2,g3,g4;
END;

(* Define the Interconnectivity of Circuit *)
NETSEGMENT;
s = .s,g3.a;
r = .r,g4.b;
cp = .cp,g3.b,g4.a;
n3 = g3.y,g1.a;
n4 = g4.y,g2.b;
q = g1.y,.q,g2.a;
qb = g1.b,g2.y,.qb;
ENDNETS;
ENDC;
```

Figure 4-2. SDL Description of RS Flip-Flop Circuit*

The structural information represented by the schematic diagram can be encoded using the SDL language, as shown in Figure 4-2. This SDL source file includes two symbol definitions, one for the nand symbol, and

---

* It is possible to automatically extract netlist information from a schematic diagram, and thus avoid the use of the direct use of the SDL language in certain implementations.

the second for the rsff symbol. The description associated with the nand symbol defines its external ports and their input/output directions, but no internal structure is given. For the rsff symbol, the external ports and input/output directions are given, as well as the internal structure of the circuit. The TYPES statement declares all symbols which will be referenced by the circuit, and is followed by a statement which defines the instance name for each NAND gate (in this case the NAND gates are named g1 through g4). The reference mechanism used in SDL allows any number of levels of hierarchy to be represented. Later examples in this chapter will illustrate how a more complex hierarchy can be used.

```
part number: SN74LS00
tPLH max: 15ns
tPLH typ: 9ns
tPHL max: 15ns
tPHL typ: 10ns
```

Figure 4-3. NAND Gate Characteristics

Behavioral information associated with each model is described using the ADLIB language. In this case, the model for a NAND gate with the characteristics shown in Figure 4-3 will be used. Normally, this type of information is extracted from a manufacturer's data book.* The ADLIB description for this NAND gate is shown in Figure 4-4. The port names shown match those in the SDL declaration for the symbol, and the names of the component type (*e.g.* nand) match the names of the SDL symbols. In this, and every other example, the names of the ports and symbols correlate the structural description for a circuit with the behavioral information.

---

*   In this case the data was taken from the Texas Instruments TTL Data Book.

```
NETTYPE
boolnet=boolean;

COMPTYPE nand;
  (* Declare Input/Output Ports *)
  INWARD a,b:boolnet;
  OUTWARD y:boolnet;

  (* Define NAND behavior *)
  SUBPROCESS
    nandfunction: TRANSMIT NOT(a AND b) CHECK a,b TO y DELAY 15;

  (* Initialize the NAND function *)
  BEGIN
  ASSIGN NOT(a AND b) to Y;
  END;
```

Figure 4-4. ADLIB Description of NAND Gate

The essence of the NAND gate's behavior is captured in the subprocess **nandfunction**, which operates as follows:

- Each time the subprocess runs, the expression NOT(a AND b) is transmitted to the output port y.

- The guard CHECK a,b determines when the expression is to be evaluated. In this case, the expression is evaluated every time the inputs a or b change value.

- The timing clause DELAY 15 specifies that the net update is to occur at the current time plus 15 nanoseconds.

The main body of the component type is used for initialization of the model. In this case, the output port y is assigned a value determined by the initial values of a and b.

Although this model captures the basic NAND function, its use is limited because it makes several simple-minded assumptions:

- Rise and fall delays are the same. In reality, rise and fall delays can differ by a large factor, especially in MOS technologies, and can also vary from one individual device to another.

- All inputs will cause a predictable output. In practice, minimum input pulse requirements must be maintained, otherwise the output can be unpredictable.

- The state of the device at power up is known. In most practical design cases, the inputs to the device, and hence the output of the device during power up, are unknown. In this model, the input values are assumed to have a determined value, and these are used to calculate an output value directly. This approach can cause initialization problems during simulation, and is a consequence of the value system chosen for this model.

The preferred solution is to use multiple logic values. For example, one could use a four-value system with the following values:

1. Low – equivalent to boolean `false`.

2. High – equivalent to boolean `true`.

3. Unknown – used during initialization and error conditions when it is not possible to determine the current value of a signal.

4. High Impedance – used to represent the high impedance state often seen in wired-gate and bus configurations.

When a value system such as this is used, the model will initialize the outputs to the unknown value, resulting in more accurate simulation results. If the circuit is designed correctly, the unknown values will usually be filtered out during power up and reset sequences, until none is left.

```
TYPE
loglevel=(unk,z,lo,hi);
NETTYPE
ttlnet=loglevel;

COMPTYPE nand(tplh,tphl,minpulse:integer);
DEFAULT tplh=15;
        tphl=15;
        minpulse=1;
INWARD a,b:ttlnet;
OUTWARD y:ttlnet;
VAR lastchange:integer;
SUBPROCESS
  constraints: UPON true CHECK a,b DO
    BEGIN
    IF time-lastchange < minpulse THEN
      writeln('Spike Detected');
    lastchange:=time;
    END;

  nandfunction: UPON true CHECK a,b DO
    BEGIN
    logprepd(lognand[a,b],tplh,tphl); (* var logretdel *)
    ASSIGN TRANSPORT lognand[a,b] TO y DELAY logretdel;
    END;
BEGIN
lastchange:=0;
END;
```

Figure 4-5. NAND Model with Delay Calculations

The ADLIB model in Figure 4-5 shows a more realistic NAND gate model which addresses these problems. In the first line of the component type definition shown, the parameters to this model are listed. These are used for passing information from the structural description file into the behavioral model on a component- by-component basis*. In this case, the rise and fall delays and the minimum pulse widths are utilized. The

---

* Most frequently, this information consists of back-annotated layout information, such as delays.

DEFAULT statement specifies default values for the parameters. As in Figure 5, the input and output port names are given along with their data types. In this case a data type of ttlnet is used to represent a four-value system for the signals. A variable lastchange is used to track the previous signal change for minimum pulse calculations. Two concurrent subprocesses, constraints and nandfunction, perform the basic processing for the model. Constraints continually watches the input stimulus and flags any input pulses that do not meet the input pulse-width requirements. An alternative approach would be to cause an unknown value to be output, or to suppress the output altogether. All approaches have shortcomings, but usually, flagging error conditions is the approach preferred by designers.

The nandfunction subprocess assigns a new output signal value whenever the inputs change. The logprepd function tests the value of a, and saves the appropriate delay value in logretdel, based on whether the new output will cause a rise or a fall. The ASSIGN TRANSPORT statement performs the same function as the previously discussed ASSIGN statement, but also suppresses any events that are scheduled after the time for the current event. This statement is required to insure that the differences between rise and fall delays do not cause faulty output given a short input pulse. Finally, the main body is used to assign the lastchange variable an initial value of zero during initialization of the model.

4.2 A High-Level Example

The following example illustrates how the ADLIB language can be used in high-level hardware design.

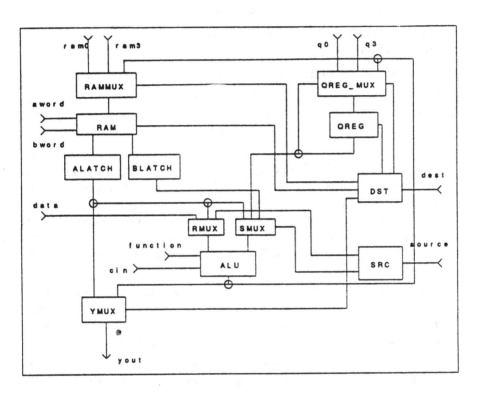

Figure 4-6. AM2901 Bit Slice Microprocessor

Figure 4-6 shows the top-level schematic diagram for the Advanced Micro Devices AM2901 bit-slice microprocessor chip, and the following list summarizes the basic functions of the inputs and outputs shown.

- Aword input − a four-bit address used to select one register whose contents are displayed through the a port of RAM.

- Bword input − a four-bit address used to select one register whose contents are displayed through the b port of RAM, and into which new data can be written when the clock goes low.

- Function input − three bits of instruction control which specify the ALU function.

- Dest input − three bits of instruction control which specify what data is to be deposited in the Q-register or the register stack.

- Source input − three bits of instruction control which specify what data sources will be applied to the ALU.

- Q3,Ram3 inputs − A shift line at the MSB of the Q-register (Q3) and the register stack (RAM3). When the destination code on dest input indicates an up shift, the tri-state outputs are enabled, the MSB of the Q-register is available on the Q3 port, and the MSB of the ALU output is available on the ram3 port. Otherwise, the tri-state outputs are off (*i.e.* high impedance). When the destination code calls for a down shift, the ports are used as the data inputs to the MSB of the Q-register and RAM.

- Q0,ram0 inputs - Shift lines like Q3 and ram3, but at the LSB of the Q-register and RAM. These ports may be tied to the Q3 and ram3 ports of an adjacent device in order to transfer data between devices for up-and-down shifts of the Q-register and ALU data.

- Data input - Direct data inputs. A four-bit data field which may be selected as one of the ALU data sources for entering data into the device. D0 is the LSB.

- Yout output - The four bits of data output. These are tri-state output lines. When enabled, they display either the four outputs of the ALU or the data on the A-port of the register stack, as determined by the destination code dest.

The reader may have noticed that several pins of the actual chip are not described in this example, *e.g.* the clock input. Since this is a high-level

---

*   As described in the Advanced Micro Devices Bipolar Microprocessor Logic and Interface Data Book for the Am2900 Family of Chips.

design example it is appropriate to ignore certain details of the circuit operation. Later in this section, the approach for handling the clock and the advantages of this abstraction will be illustrated.

```
NETTYPE
  (* two value logic system *)
  two_value = boolean;
  (* three value logic system *)
  three_value = (low,high,z);
  (* 4 bit data bus *)
  data_path = 0..15;
  (* source control values *)
  source = (aq,ab,zq,zb,za,da,dq,dz);
  (* ALU control values *)
  alu_function = (fadd,fsubr,fsubs,flor,fand,
                  fnotrs,fexor,fexnor);
  (* RAM and QREG multiplexer controls *)
  mux_control = (shift_up,shift_down,no_shift);
  (* RAM control *)
  ram_control = (read_ram,write_ram);
  (* Reg enable *)
  reg_enable = (enable,disable);
  (* Y output MUX *)
  ymux_control = (aoutput,youtput);
  (* Destination decoder control *)
  destination = (qrg,nop,rama,ramf,ramqd,ramd,ramqu,ramu);
  (* RMUX control *)
  rmux_control = (rmux_a,rmux_d,rmux_inhibit);
  (* SMUX control *)
  smux_control = (smux_a,smux_b,smux_q,smux_inhibit);
```

Figure 4-7. AM2901 Net Type Definitions

One of the most important capabilities of the ADLIB language is the ability to establish abstract data types for the signals and busses in a circuit. In this case, a number of different net types are established, as shown in Figure 4-7. Most of the busses shown in Figure 4-6 represent multi-bit signals.

I5	I4	I3	Octal	Mnemonic	Function
L	L	L	0	ADD	R
L	L	H	1	SUBR	S
L	H	L	2	SUBS	R
L	H	H	3	OR	R
H	L	L	4	AND	R
H	L	H	5	NOTRS	RNOT
H	H	L	6	EXOR	R
H	H	H	7	EXNOR	R

Figure 4-8. ALU Function Bit Encoding

All busses use a bit-encoding scheme to represent instructions or operations. For example, Figure 4-8 shows the bit encoding scheme that is used, in the case of the three-bit input called funct. Through the use of an abstract net-type definition, the semantics of a collection of signals such as those of the funct bus can be represented. The net-type definition alu_function, as shown in Figure 4-8, gives the ADLIB definition for the bus in this example.

Synchronous Activity of the AM2901

Many electronic systems operate synchronously, which is why the ADLIB language has a built-in capability for modeling synchronous behavior. In this example, a clock called clk is defined as:

```
(*latch/reg clock, phase 1=high, phase 2=low *)
CLOCK clk (100,2);
```

Several of the models described later in this section will utilize this clock definition to coordinate their activities. The clock defined here has two phases, each with a 100 nanosecond duration.

```
PACKAGE am2901;
  FUNCTION shift(upval,downval,crntval:integer;up:boolean):
   integer;
  (* shift up or down crntval using upval or downval*)
   BEGIN
   (* watch out for Z values *)
   upval:=upval MOD 2;
   downval:=downval MOD 2;
   IF up THEN
     BEGIN
     crntval:=(crntval*2) MOD 15;
     crntval:=crntval+upval;
     END
   ELSE crntval:=(crntval DIV 2) + (downval*8);
   shift:=crntval;
   END;
(* package body *)
BEGIN
END.
```

Figure 4-9. AM2901 Support Package

PACKAGEs for the AM2901

In ADLIB it is possible to define a *package*, which is a collection of subroutines that will be use by the various component types in design. Figure 4-9 shows the definition for the AM2901 package. In this case, only one subroutine is defined, called shift, and the body for the package performs no function. The shift subroutine will be used later in this section to shift four-bit words up and down.

```
COMPTYPE ram_mux;
  INWARD y:data_path;
         cntrl:mux_control;
  OUTWARD d:data_path;
  BOTHWAYS ram0,ram3:three_value;
  SUBPROCESS
    multiplex: UPON true CHECK cntrl,y DO
      BEGIN
      CASE cntrl OF
        shift_up:
          BEGIN (* shift last bit to RAM3 *)
          IF y>=8 THEN ASSIGN high TO ram3
          ELSE ASSIGN low TO ram3;
          (* read shift in from RAM0 *)
          ASSIGN shift(ORD(ram0),ORD(ram3),y,true) TO d;
          ASSIGN z TO ram0; ASSIGN z TO ram3;
          END;
        shift_down:
          BEGIN (* shift last bit to RAM0 *)
          IF (y MOD 2) = 1 THEN ASSIGN high TO ram0
          ELSE ASSIGN low TO ram0;
          (* read shift in from RAM3 *)
          ASSIGN shift(ORD(ram0),ORD(ram3),y,false) TO d;
          ASSIGN z TO ram0; ASSIGN z TO ram3;
          END;
        no_shift:
          BEGIN (* don't use the RAM shift inputs *)
          ASSIGN z TO ram0; ASSIGN z TO ram3;
          ASSIGN y TO d;
          END;
        END;
  BEGIN
  ASSIGN 0 TO d;
  ASSIGN z TO ram0; ASSIGN z TO ram3;
  END;
```

Figure 4-10. RAM Multiplexer Model

The RAM Multiplexer for the AM2901

Figure 4-10 shows the model for the RAM multiplexer. This block feeds the RAM with one of three possible inputs:

- shift_up - Shift the MSB of Y input to ram3. Shift the LSB in from ram0. Pass the results on to the RAM.

- shift_down - Shift the MSB in from ram3. Shift the LSB out from ram0. Pass the results on to the RAM.

- No_shift - Ignore the ram0 and ram3 signals. Pass the Y input directly to RAM.

This component type illustrates several aspects of model writing in ADLIB. First, the ram0 and ram3 signals are bidirectional, since they are multi-sourced signals. The next chapter will discuss in more detail how busses and wired-gate situations are handled in ADLIB. In this particular case, whenever the model is not driving these signals a high-impedance value is output to indicate that the bus is not being driven by the multiplexer. The shift subroutine, taken from the previously described PACKAGE am2901 is used to perform the shift operations for this model. (See Figure 4-9 for details on how this function operates.) Finally, the body of the RAM multiplexer model is used to set initial values to the D, RAM0, and RAM3 signals.

```
COMPTYPE qreg_mux;
  INWARD cntrl:mux_control;
         y,q:data_path;
  OUTWARD dout:data_path;
  BOTHWAYS q0,q3:three_value;
  SUBPROCESS
    (* watch for changes in cntrl or inputs
     *** and update outputs *)
    multiplex: UPON true CHECK cntrl,y,q DO
      BEGIN
      CASE cntrl OF
        shift_up:
          BEGIN
          (* shift last bit to Q3 *)
          IF q>=8 THEN ASSIGN high TO q3
          ELSE ASSIGN low TO q3;
          (* read shift in from q0 *)
          ASSIGN shift(ORD(q0),ORD(q3),q,true)
            TO dout;
          ASSIGN z TO q0; ASSIGN z TO q3;
          END;
        shift_down:
                      ...
        no_shift:
          BEGIN
          (* don't use Q0, Q3...put ALU output on Q *)
          ASSIGN z TO q0; ASSIGN z TO q3; ASSIGN y TO dout;
          END;
        END;
      END;
    BEGIN
    ASSIGN z TO q0; ASSIGN z TO q3; ASSIGN 0 TO dout;
    END;
```

Figure 4-11. Q Register Multiplexer

Q Register Model

Figure 4-11 shows the model for the Q register multiplexer. This model is similar to the previously discussed ALU Multiplexer and demonstrates similar features. The only major difference in operation is that during shift operations the Q input is used, while for non-shift operation the Y input is used.

```
COMPTYPE ram; (* 16 word by 4 bit 2-port RAM *)
  INWARD b_addr,a_addr,d:data_path;
        cntrl:ram_control;
  OUTWARD a,b:data_path;
  VAR memory:ARRAY[data_path]OF data_path;
      i:integer; write_phase:boolean;
  SUBPROCESS
    write_mode: UPON true SYNC clk PHASE 0 DO
      BEGIN (* watch for write mode *)
      IF flag(1) THEN writeln('-RAM- write mode');
      write_phase:=true;
      IF cntrl = write_ram THEN
        BEGIN
        memory[b_addr]:=d;
        ASSIGN d TO b;
        END;
      END;
    read_mode:UPON true SYNC clk PHASE 1 DO
      write_phase:=false;
    ram_control: UPON true CHECK cntrl,b_addr,a_addr,d DO
      BEGIN(* watch for changes in RAM read/write control or inputs *)
      IF cntrl=read_ram THEN
        BEGIN (* change to read mode *)
        ASSIGN memory[a_addr] TO a; ASSIGN memory[b_addr] TO b;
        END
      ELSE
        BEGIN
        ASSIGN memory[a_addr] TO a;
        IF write_phase THEN
          BEGIN
          memory[b_addr]:=d; ASSIGN d TO b;
          END;
        END;
      END;
  BEGIN
  write_phase:=true;
  FOR i:=0 to 15 DO memory[i]:=0;
  ASSIGN 0 TO a; ASSIGN 0 TO b;
  END;
```

Figure 4-12. Random Access Memory Model

RAM Model

Figure 4-12 shows the model for the RAM for this circuit. The operation
of this block can be described as follows:

• Data in any of the 16 words (four bits each) can be read from the a
  port via the a_addr address input.

- Data in any of the 16 words can also be simultaneously read from the b port via the `b_addr` address input.

- When the write mode is enabled, new data is always written to the address specified by the `b_addr` input.

In this model a local variable `memory` is declared which is used to store the state of the model. This model operates synchronously, changing state only when the clock `clk` changes its phase. A second local variable called `write_phase` is used to retain the read-or-write mode of the RAM.

The `write_mode` subprocess is synchronized to phase 0 of the clock, and a write will occur if the RAM write control is enabled. The `read_mode` subprocess operates during the other phase of the clock, and the `ram_control` subprocess operates asynchronously whenever any of the inputs to the RAM change values.

Several `writeln` statements are included to track operation of the RAM, and, in addition, the `flag` function is used for conditional tracking. The argument, 1, specifies that flag number 1 must be enabled in order for the function to return `true`. These flag values are controlled by control commands during simulation.

```
COMPTYPE qreg;
  INWARD d:data_path;
         cntrl:reg_enable;
  OUTWARD q:data_path;
  SUBPROCESS
    (* output value changes on low to high transition*)
    clk_output: UPON cntrl = enable SYNC clk PHASE 1 DO
      ASSIGN d TO q;
  BEGIN
  ASSIGN 0 TO q;
  END;
```

Figure 4-13. Q Register Model

The model for the Q Register shown in Figure 4-13 is a relatively simple model. It propagates its input to its output, synchronized to clock phase 1, and operates only only when it is enabled. The state of the register is stored in the output variable and no additional memory is required.

```
COMPTYPE latch;
  INWARD inv:data_path;
  OUTWARD outv:data_path;
  VAR pass_phase:boolean;
  SUBPROCESS
    pass_mode1:UPON true SYNC clk PHASE 0 DO
      pass_phase:=false;
    pass_mode2:UPON true SYNC clk PHASE 1 DO
      BEGIN
      pass_phase:=true;
      ASSIGN inv TO outv;
      END;
    clk_output:UPON pass_phase CHECK inv DO
      ASSIGN inv TO outv;
  BEGIN
  pass_phase:=false;
  ASSIGN 0 TO outv;
  END;
```

Figure 4-14. Latch Model

Figure 4-14 shows the model for the a and b latches. The model passes the input value directly to the output while it is in the pass phase (as determined by the pass_phase variable), and retains the last clocked value while not in this mode. The two subprocesses pass_mode1 and pass_mode2 are used to update the mode of the latch, and the subprocess clk_output is active during the pass-through mode to update the output asynchronously whenever the input changes value.

```
COMPTYPE dest_decoder;
  INWARD dest:destination;
  OUTWARD qmux,rammux:mux_control;
          ram:ram_control;
          qreg:reg_enable;
          ymux:ymux_control:
  SUBPROCESS
    proc_destination:UPON true CHECK dest DO
      BEGIN
      CASE dest OF
        qrg: (* F -> Q *)
          BEGIN
          ASSIGN read_ram TO ram;
          ASSIGN no_shift TO qmux;
          ASSIGN enable TO qreg;
          ASSIGN youtput TO ymux;
          END;
            ...
        ramu: (* 2F -> B *)
          BEGIN
          ASSIGN shift_up TO rammux;
          ASSIGN write_ram TO ram;
          ASSIGN disable TO qreg;
          ASSIGN youtput TO ymux;
          END;
        END;
      END;
  BEGIN
  ASSIGN read_ram TO ram; ASSIGN disable TO qreg;
  ASSIGN no_shift TO rammux; ASSIGN no_shift TO qmux;
  ASSIGN youtput TO ymux;
  END;
```

Figure 4-15. Destination Decoder

The destination decoder is shown in Figure 4-15 (part was omitted for brevity). Operation in this model is stimulated by any change in the input destination signal dest. Mneumonic values for destination categories (*e.g.* qrg, rama, *etc.*) have been used to increase the readability in this model. In this case, the values chosen match the microcode, making interpretation of this model easier than if binary or hex values had been used.

```
COMPTYPE rmux;
  INWARD d,a:data_path;
        cntrl:rmux_control;
  OUTWARD aout:data_path;
  SUBPROCESS
    rmux_output:UPON true CHECK cntrl,d,a DO
      IF cntrl=rmux_a THEN ASSIGN a TO aout
      ELSE IF cntrl=rmux_d THEN ASSIGN d TO aout
      ELSE ASSIGN 0 TO aout;
  BEGIN
  ASSIGN 0 TO aout;
  END;
```

Figure 4-16. Mux Model

The r, s and y multiplexers are roughly equivalent in their operation, but differ in the number of inputs they handle. Figure 4-16 shows a model for the rmux.

```
COMPTYPE alu;
  INWARD func:alu_function;
         r,s:data_path;
         cin:two_value;
  OUTWARD y:data_path;
          cout:two_value;
  VAR work:integer;
  SUBPROCESS
    alu_output:UPON true CHECK func,r,s,cin DO
      CASE func OF
        fadd:   BEGIN
                IF cin THEN work:=r+s+1 ELSE work:=r+s;
                ASSIGN work>15 TO cout;
                ASSIGN work MOD 15 TO y;
                END;
        fsubr:  BEGIN
                IF cin THEN work:=s-r-1 ELSE work:=s-r;
                ASSIGN work<-15 TO cout;
                ASSIGN work MOD 15 TO y;
                END;
                     ...
        END;
  BEGIN
  ASSIGN 0 TO y; ASSIGN false TO cout;
  END;
```

Figure 4-17. ALU Model

Figure 4-17 shows the ALU model (six cases were omitted for brevity).
The ALU operates asynchronously, updating its outputs whenever the
inputs change, through the use of subprocess alu_output. Again,
readability is improved by using mneumonic values for the ALU control.

```
COMPTYPE source_decoder;
  INWARD source:source;
  OUTWARD smux:smux_control;
          rmux:rmux_control;
  SUBPROCESS
    source_control: UPON true CHECK source DO
      CASE source OF
        aq: BEGIN (* a and q to alu *)
            ASSIGN rmux_a TO rmux;
            ASSIGN smux_q TO smux;
            END;
        ab: BEGIN (* a and b to alu *)
            ASSIGN rmux_a TO rmux;
            ASSIGN smux_b TO smux;
            END;
        END;
BEGIN
ASSIGN rmux_inhibit TO rmux;
ASSIGN smux_inhibit TO smux;
END;
```

Figure 4-18. Source Decoder Model

Finally, Figure 4-18 shows the model for the source decoder (minus six cases). This model is similar to the destination decoder shown in Figure 4-15.

4.3 Modeling Software

The following example illustrates how the ADLIB language can be used for software analysis. In this case, an algorithm for the data-link layer protocol[TAN81] of a point-to-point local area network will be analyzed to determine the following:

- Will the algorithm work correctly under all conditions?

- Will the algorithm meet the performance expectations?

- What are the design requirements for network nodes (buffer sizes, time-out periods, *etc.*)?

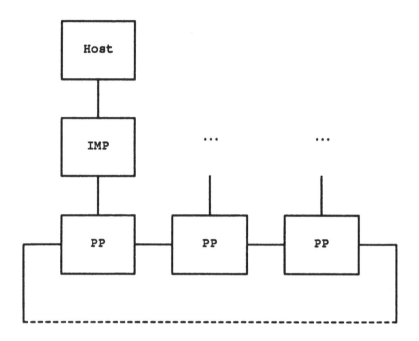

Figure 4-19. Network and Protocol Layering

Figure 19 shows the relationship between the various nodes in this point-to-point network, and the network handling protocol layering used. In this case, it is assumed that the Physical Protocol box (PP) will handle all aspects of transmitting and receiving raw bits over the network, including check-sum processing. In addition, the study will limit itself to the data-link layer (referred to as the "data link service data unit" from ISO terminology [ZIM80]) of processing by the Interface Message Processor (IMP), in which the network is made to appear to the host to be free of transmission errors. This task is accomplished by breaking up the input data into data frames, transmitting the frames sequentially, and processing the acknowledgement frames sent back by the receiver.

The following summarizes the functions that the algorithms must perform:

- It must transmit and receive host messages between other nodes in network and the local node.

- It must manage handshaking with other network nodes when data is being sent or received.

- It must hide transmission errors from the host by forcing retransmission of data when necessary.

This example implements a sliding-window protocol in which nonsequential receives from another node are allowed. The following list summarizes the algorithm's operation:

- The receiving IMP must send an ACK to the sender to acknowledge message.

- The receiving IMP must send a NAK to the sender to signal improper message, and request a resend.

- Messages are allowed to be transmitted out of sequence. Coordination is achieved through use of sequence numbers in a sliding-window protocol.

- ACK's are piggy-backed onto messages where possible, *i.e.* when outgoing message occurs within a given time limit.

- Messages are retransmitted after a given time limit if no ACK is received.

```
TYPE
   eventtype=(nulevent, framearrival, cksumerr,
             timeout, hostready, hostidle);
   framekind=(dataframe, ackframe, nakframe);
   message=ARRAY[0..lastbit]OF boolean;
   frametype=RECORD
     kind:framekind;
     seq:integer;
     ack:integer;
     info:message;
     END;
   bufnr=0..maxbuf;
   sequencenr=0..maxseq;
NETTYPE
   frame=frametype;
   eventnet=eventtype;
   timernum=integer;
```

Figure 4-20. Network Net Types

```
COMPTYPE host;
  BOTHWAYS port:frame;
  VAR fbuff:frame;
  PROCEDURE fromhost(VAR info:message);
  (* return randomly generated message *)
    VAR i:integer;
    BEGIN
    (* set up random bit string to
      ** represent host message *)
    FOR i:=0 TO lastbit DO info[i]:=rnddraw(0.5);
    END;
  SUBPROCESS
    sndmessage:UPON true DELAY rndint(msgmin,msgmax) DO
      BEGIN
      fromhost(lbuff.info);
      ASSIGN lbuff TO port;
      END;
  BEGIN
  END;
```

Figure 4-21. Host Model

For the remainder of this section the declarations shown in Figure 4-20 will be used. Figure 4-21 shows the ADLIB source for the host box model (see Figure 4-19). In this example, the subprocess sndmessage initiates new messages, which are generated by the fromhost subroutine at random intervals. These messages come with a minimum time delay of msgmin and a maximum delay of msgmax. This type of stochastic modeling is most useful during the early stages of design, when only performance specifications and rough models of the surrounding system are available.

```
COMPTYPE imp;
  BOTHWAYS hostport,netport:frame;
  INTERNAL timer:timernum;
          idle:timernum;
  VAR ackexpected:sequencenr;
       (* lwr edge of trns wind *)
      nextframetosend:sequencenr;
       (* upp edge of trns wind *)
      frameexpected:sequencenr;
       (* lwr edge of rcv wind *)
      toofar:sequencenr;
       (* upp edge of rcv wind *)
      oldestframe:sequencenr;
       (* which frame time out? *)
      i:bufnr; (* indx into buffer pool *)
      s:frame; (* scratch variable *)
      outbuf:ARRAY[bufnr]OF msgpnt;
       (* bfrs/outbound streams *)
      inbuf:ARRAY[bufnr]OF msgpnt;
       (* bfrs/inbound streams *)
      arrived:ARRAY[bufnr]OF boolean;
       (* inbound bit map *)
      nbuffered:sequencenr;
       (* # bfrs currently used *)
      nonak:boolean;(* false when see a nak *)
      savetime:ARRAY[bufnr]
        OF utl_evtype;(* remove interrupts *)
      saveidle:utl_evtype;   (* save ack time-outs *)
       ...
      (* support subroutines/models shown later *)
       ...
  BEGIN
  END;
```

Figure 4-22. IMP Model

Figure 4-22 shows the declaration sections for the IMP box model (see Figure 4-19) in which the data-flow algorithm is described. This code declares the input/output ports hostport and netport, two internal input/output nets timer and idle, and the local variables (state) for the model. The two internal nets act in exactly the same way an external port would, except that they are strictly local to the model. Conceptually, they correspond to a signal which resides entirely within the model. In this case these two internal nets are used for interrupt processing.

Several possible interrupts can occur, described as follows:

- Frame Arrival (Framearrival) − a frame has arrived via the network.

- Check Sum Error (Cksumerr) − a frame has arrived via the network, but it is damaged.

- ACK Time-out (timeout) − the IMP has received no ACK for a message recently transmitted on the network, and the ACK interrupt handler has timed out. This interrupt subprocess issued a signal that the maximum allowed wait for an ACK frame has been exceeded.

- Host Message (Hostready) − a message send request has been received from the host.

- Piggy-back Time-Out (hostidle) − the piggy-back interrupt handler has timed-out. This interrupt subprocess is used to signal that the maximum allowable delay for piggy-backing an ACK has been exceeded, and the algorithm should transmit an ACK immediately.

```
getframe:UPON true CHECK netport DO (*network has sent message*)
  BEGIN
  IF rnddraw(0.1) THEN (* 10% chance of check-sum error *)
    BEGIN
    IF nonak THEN sendframe(nakframe,0);
    END
  ELSE (* process frame received from network *)
    BEGIN
    IF netport.kind=dataframe THEN
      BEGIN (* an undamaged data frame has arrived *)
      IF (netport.seq<>frameexpected)AND nonak THEN
        sendframe(nakframe,0);
      (* frames may be accepted in any order *)
      IF between(frameexpected, netport.seq,toofar) AND
         (arrived[netport.seq MOD nrbufs]=false) THEN
        BEGIN
        arrived[netport.seq MOD nrbufs]:=true;(*mark buffer full*)
        (* insert data in buffer *)
        inbuf[netport.seq MOD nrbufs]:=netport.info;
        (* pass frames and advance window *)
        WHILE arrived[frameexpected MOD nrbufs] DO
          BEGIN
          (* pass info to host *)
          s.info:=inbuf[frameexpected MOD nrbufs];
          ASSIGN s TO hostport;
          nonak:=true;
          arrived[frameexpected MOD nrbufs]:=false;
          inc(frameexpected); (*advance low edge of recv. window*)
          (* advance upper edge+1 of receiver window *)
          inc(toofar);
          startacktimer; (* to see if separate ack needed *)
          END;
        END;
      END;
    (* re-transmit frame if NAK seen *)
    IF (netport.kind=nakframe) AND
       between(ackexpected,(netport.ack+1)mod(maxseq+1),
               nextframetosend) THEN
      sendframe(dataframe,(netport.ack+1)mod(maxseq+1));
    (* disable ACK timers for all valid frames *)
    WHILE between(ackexpected,netport.ack,nextframetosend) DO
      BEGIN
      nbuffered:=nbuffered-1; (* handle piggy-backed ack *)
      stoptimer(ackexpected MOD nrbufs); (* frame arrived intact *)
      inc(ackexpected); (* advance lower window edge *)
      END;
    END;
  END;
```

Figure 4-23. Frame Handling Subprocess

For each of these possibilities a concurrent subprocess is described which
will handle the interrupt. The frame arrival processing code and and
check-sum error processing code is shown in Figure 4-23. This subprocess
will handle incoming frames from the network. A 10% chance of check-
sum errors is introduced by this subroutine to model the possibility of

transmission errors. In addition, this subroutine will verify that all undamaged frames have the correct sequence number, and will process out-of-order frames using the sliding-window protocol.

```
(* timer for receiving message acknowledgement *)
timercheck:UPON true CHECK timer DO
  BEGIN
  IF flag(1) THEN dumpstat;  (* debug trace *)
  sendframe(dataframe,oldestframe);
  END;
```

Figure 4-24. ACK Time-Out Handling Subprocess

```
PROCEDURE starttimer(k:sequencenr);
  BEGIN
  ASSIGN k TO timer DELAY timeoutdelay;
  utlsaveev(savetime[k]);
  END;

PROCEDURE stoptimer(k:sequencenr);
  BEGIN
  IF savetime[k].utime>time THEN utldelev(savetime[k]);
  END;
```
Figure 4-25. ACK Interrupt Support Subroutines

The code for processing acknowledgement time-outs is shown in Figure 4-24. This subroutine will process time-out interrupts when an ACK is not received. Two support subroutines are associated with this subprocess. The starttimer subroutine is called to initiate a time-out delay for a message ACK, (which is associated with a particular message sequence-number). The stoptimer subroutine is called to cancel the ACK time-out. Both these subroutines are shown in Figure 4-25. In essence, the starttimer subroutine schedules interrupts for the future, and the stoptimer subroutine de-schedules a given interrupt using the utldelev utility.

```
(* host initiates message *)
host:UPON true CHECK hostport DO
  BEGIN
  IF flag(1) THEN dumpstat;  (* debug trace *)
  (* expand window *)
  nbuffered:=nbuffered+1;
  (* fetch new message from host *)
  outbuf[nextframetosend MOD nrbufs]:=hostport.info;
  (* transmit frame *)
  sendframe(dataframe,nextframetosend);
  (* advance upper window edge *)
  inc(nextframetosend);
  END;
```

Figure 4-26. Host Message Handling Subprocess

The code to process messages from the host is shown in Figure 4-26. This code saves the message in an output buffer `outbuf`, and transmits the frame onto the network.

```
(* waiting for piggy-back message for ACK *)
idlehost: UPON true CHECK idle DO
  BEGIN
  IF flag(1) THEN dumpstat;  (* debug trace *)
  sendframe(ackframe,0);
  END;
```

Figure 4-27. Piggy-Back Time-Out Handling Subprocess

The code to handle piggy-back acknowledgements is shown in Figure 4-27. This code will process time-out interrupts when no message request is made from the host for a maximum time period, thus forcing the IMP to send an ACK message since it cannot piggy-back the ACK.

```
PROCEDURE startacktimer;
  BEGIN
  ASSIGN (idle+1) MOD 2 TO idle DELAY hostdelay;
  utlsaveev(saveidle);
  END;

PROCEDURE stopacktimer;
  BEGIN
  IF saveidle.utime>time THEN utldelev(saveidle);
  END;
```

Figure 4-28. Piggy-Back Interrupt Support Subroutines

Two support subroutines are associated with this subprocess. The startacktimer subroutine is called to initiate a time-out delay for a message ACK (associated with a particular message sequence number). The stopacktimer subroutine is called to cancel the ACK time-out. Both these subroutines are shown in Figure 4-28. In essence, the startacktimer subroutine schedules interrupts for the future, and the stopacktimer subroutine de-schedules a given interrupt.

```
FUNCTION between(a,b,c:sequencenr):boolean;
  VAR ret:boolean;
  BEGIN
  IF ((a<=b)AND(b<c))OR
     ((c<a)AND(a<=b))OR
     ((b<c)AND(c<a)) THEN between:=true
  ELSE between:=false;
  END;


PROCEDURE sendframe(fk:framekind;framenr:sequencenr);
(* construct and send a data, ack, or nak frame *)
  BEGIN
  (* kind = data, ack or nak *)
  s.kind:=fk;
  IF fk=dataframe THEN
   s.info:=outbuf[framenr MOD nrbufs];
  s.seq:=framenr;  (* only meaningful for data frames *)
  s.ack:=(frameexpected+maxseq) MOD (maxseq+1);
  (* one nak per frame, please *)
  IF fk=nakframe THEN nonak:=false;
  (* transmit the frame *)
  ASSIGN s TO netport;
  IF fk=dataframe THEN starttimer(framenr MOD nrbufs);
  stopacktimer;
  END;
```

Figure 4-29. Frame Support Subroutines

Associated with the above subprocesses are the two other support subroutines shown in Figure 4-29. The between function is used to determine whether a valid sequence number has been seen. The sendframe subroutine is used to transmit frames to the network.

4.4 Protocol Example Summary: Using ADLIB for Algorithm Debugging

This section has shown how a network protocol algorithm can be described. The simulation features which support this prototyping environment are:

- Support of "black-box" modeling with concurrency and inter-process communication.

- Access to software packages, such as the stochastic modeling primitives rndint and rnddraw used in the code samples.

- Built-in event scheduling through the language primitives, and de-scheduling through support routines.

- The ability to intermix models of hardware with models of software.

These features help a designer to study problems without having to implement a large set of support subroutines. In this particular case, a designer was able to develop the entire data-link-layer algorithm in less than one day.

```
----------HOSTB----------
AckExpected: 0
NextFrameToSend: 0
FrameExpected: 0
TooFar: 3
OldestFrame: 8224
Arrived: 0 [0]:F [1]:F [2]:F
NBuffered: 0
NoNak: True
HOSTB next event: HOSTREADY
Host HOSTB transmitted: 11000001110
IMP HOSTB sending frame: <DATA seq:0 ack:5 11000001110>
HOSTB initiating receiver timeout for sequence 0
  for time 331
HOSTB removing piggy-back timeout
```

Figure 4-30. Model Debugging

4.5 Model and Algorithm Debugging

At this point it is appropriate to discuss the interactive simulation of this network. The actual interface varies from one implementation to another, but it is always possible to include Pascal statements to display and control the simulation. For this example, Figure 4-30 shows the simulation output. Overall performance of the network can also be analyzed by collecting more data and producing statistics and graphics (the Appendix illustrates this).

Independently of the source code debugging output, in some implementations the simulator is able to monitor signal activity, which allows the user to determine the following information quickly:

- The total number of frames transmitted over a given period of time

- The number of acknowledgement frames transmitted over a given period of time
- The number of negative acknowledgement frames transmitted over a given period of time
- The number of piggy-backed frames transmitted over a given period of time
- The number of check-sum errors encountered.

Information Produced by ADLIB simulation

In summary, the ADLIB environment helps the designer to assess many of the critical aspects of design early in the design process, including questions such as:

- Will the algorithm work correctly under all conditions?
- Will the algorithm meet performance expectations?
- What are the resource requirements for network nodes (buffer sizes, time-out periods, *etc*)?.

4.6 An Hierarchical Example

This section will illustrate how the various levels of design are tied together using the ADLIB language. It is tied to Figure 4-6 which shows the block-level schematic diagram for the AM2901 microprocessor chip. This section will illustrate how the ADLIB language is used to describe and simulate the ALU part of this circuit, and will follow this hierarchical design down to the gate level.

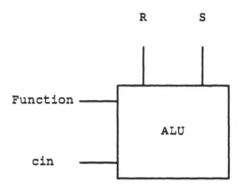

Figure 4-31. ALU Symbol

Figure 4-31 shows the symbolic representation of the ALU block. Recall Figure 4-17 in which the behavioral model for this block was given. This behavioral model can be considered to be a high-level specification for the desired operation of the ALU block. During the simulations shown earlier, this specification was simulated and debugged. In the following examples, the implementation for this ALU block will be developed. The ADLIB language allows the designer to compare the simulation results of the implementation of the ALU block back to the specification to insure that no discrepancies exist. This approach results in a less error prone design than is possible without computer validation.

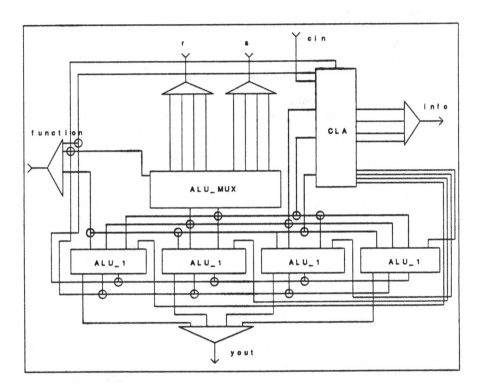

Figure 4-32. ALU Schematic Diagram

Figure 4-32 shows the schematic diagram for the ALU block. Four one-bit ALU blocks are seen, along with a multiplexer and carry-lookahead block. In addition five triangular symbols are shown. These are special "pseudo-components" called *translators* which allow mixing of simulation abstraction levels.

```
COMPTYPE salu(pdelay:INTEGER);
  DEFAULT pdelay=30;
  INWARD s0,s1,s2,s3,srs,m,cn,r,s:net_logic;
  OUTWARD fout:net_logic;
  VAR exor:logtype;
  SUBPROCESS
    calc:UPON TRUE CHECK s0,s1,s2,s3,cn,r,s,srs,m DO
      BEGIN
      exor:=logor[logand[r,lognot[s]],logand[lognot[r],s]];
      CASE logencode8(lo,lo,srs,m,s3,s2,s1,s0) OF
        27: ASSIGN logor[r,s] TO fout DELAY pdelay;
        30: ASSIGN logand[r,s] TO fout DELAY pdelay;
        24: ASSIGN logand[net[r],s] TO fout DELAY pdelay;
        25: ASSIGN exor TO fout DELAY pdelay;
        22: ASSIGN lognot[exor] TO fout DELAY pdelay;
        9: ASSIGN logexor[r,logexor[s,cn]] TO fout DELAY pdelay;
        40: ASSIGN logexor[lognot[r],logexor[s,lognot[cn]]]
              TO fout DELAY pdelay;
        8: ASSIGN logexor[r,logexor[lognot[s],lognot[cn]]]
              TO fout DELAY pdelay;
      END;
  BEGIN
  END;
```

Figure 4-33. ADLIB Model for One-Bit ALU

Each of the blocks shown in Figure 4-32 will typically have a behavioral description written in ADLIB. For example, Figure 4-33 shows the ADLIB model for the one-bit ALU. A similar model exists for both the ALU multiplexer and the carry-lookahead block. Simulation at this level allows the designer to debug the ALU design without worrying about the implementation details of the one-bit ALU components. This approach allows quicker debugging of the design because each circuit under study is of manageable size. In addition, changes can be made without concern about lower-level circuitry.

Once the designer is satisfied with simulation of the ALU at this level, he or she can proceed with implementation of the one-bit ALU, the ALU multiplexer and the carry-lookahead blocks. This top-down design methodology is continued until the entire system is implemented using primitive components, which may include gate-array macros, standard cells, or even individual transistors.

4.7 Summary — Multi-Level Simulation

The ALU input function port has a user-defined data type named funct; its binary representation was given in Figure 4-8. To reflect the gate-level schematic representation of the ALU block, it is necessary to *translate* from the higher level funct data type into the lower-level logic data type. In the schematic, a triangular translator symbol is used to indicate where this occurs, and the code for it is shown in the Appendix. In the ADLIB language, translators are defined using exactly the same syntax and semantics as component-type definitions, except that they are introduced by the keyword TRANSLATOR instead of the keyword COMPTYPE. From the standpoint of the actual circuit behavior these translators perform no function, and in fact, are removed automatically when the structural information is used in circuit fabrication. However, from the standpoint of simulation, these translators are important since they allow a single simulation run to mix levels of abstraction. This capability is one of the most important in the ADLIB language, and the following list summarizes its advantages:

- Low-level simulation of a circuit can proceed even though parts of a circuit are not fully described. For example, gate-level simulation of the ALU can be performed even though all other block functions are specified only at the top level.

- Low-level simulations of portions of a circuit can be compared to the high-level specification for these same portions. For example, one can simulate the AM2901 once using the ADLIB model for the ALU, and a second time using the gate-level implementation. Comparison of these two simulation runs make it possible to determine whether there are discrepancies between the specification for the ALU and its implementation.

- Test stimulus can be given in high-level symbolic form even for use with low-level simulations. In this example, stimulus to the ALU block using the abstract funct net type is used, even though the ALU block may be simulated at the gate level. In essence, the system is performing automatic assembly of the symbolic instructions into a binary representation.

- Finally, different levels of detail can be used to represent the same signal. For example, a boolean signal from a high-level component model can be used to drive a multi-value net, such as ttlnet described earlier in this chapter. When translating from a low-level net type to a higher level net type, information must, of necessity, be thrown away. Likewise, in translating from a high-level net type to a low-level one, information must be "synthesized," or approximated. This is the basis of the definition of multi-level simulation, as given in Chapter 2, as a *abstract approximation* − information at high levels approximates the detailed data. An example of such a multi-value/ boolean translator is shown in the Appendix, and more discussion is available in the original thesis [HIL80].

It is exactly this process of abstraction and approximation that makes the multi-level design methodology efficient, both in terms of human effort and computer resources.

CHAPTER 5

ADVANCED PRINCIPLES

This chapter contains advanced modeling examples that demonstrate the full power of the ADLIB language. Included are examples that show the how ADLIB can be used to support special types of simulation, including timing verification, fault simulation and symbolic simulation.

5.1 Using Model Main Bodies

In several of the following examples, the main body of the model is used to perform the bulk of the processing. This approach differs from examples cited previously in which the main body was used primarily for initialization and the actual processing was performed by one or more subprocesses. There are two reasons for using the main body of the model for behavioral modeling:

- With many models only one process is required; eliminating the need for multiple concurrent subprocesses.

- Using the main body tends to be more efficient at simulation run-time. As will be explained in Chapter 6, this is because it does not require as much overhead in the expanded Pascal code. On the other hand, although efficiency may be improved by using the main body of a model to control its actions, readability may be degraded. For this reason, only specialized models, which are often written by CAD experts for special purposes, normally make use of this technique.

The use of the main body to control an ADLIB model of a NAND gate is illustrated here:

```
COMPTYPE nand;
  INWARD a,b:boolnet;
  OUTWARD y:boolnet;
  BEGIN
  (* make model sensitive to a,b inputs *)
  SENSITIZE(a,b);
  (* loop forever *)
  REPEAT
    (* sleep until a or b change value *)
    DETACH;
    (* perform nand function *)
    ASSIGN NOT(a AND b) TO y DELAY 10;
  UNTIL false;
  END;
```

Figure 5-1. NAND Gate Using the Main Body for Control

This model can be contrasted to that described in the previous chapter in which another model of a NAND gate was described using the subprocess approach. The **sensitize** statement is used to establish one or more ports to which the model will be sensitive. The **desensitize** statement is used to negate the function of the **sensitize** statement. These statements can be used to change the set of sensitized ports dynamically. The **DETACH** statement causes the model to sleep until one of the sensitized ports changes value. When a model is described that makes use of the main body of a component type, a infinite loop must be established. When this loop is combined with the **sensitize** and the **DETACH** statements, a working model is created. In the example shown here, the model is sensitive to the **a** and **b** inputs, and sleeps, due to **DETACH** until either or both of these inputs changes value. Each time it is activated the model will assign the NAND function to the output port and then return to sleep immediately.

5.2 Driver Models

An ADLIB *driver model* is used to describe the environment which surrounds a given circuit being simulated. A typical use of a driver model would be in modeling the behavior of an application environment for a microprocessor. For instance, a statistical model for some proposed I/O controllers might be used in the driver model to test the microprocessor circuit.

The following rules are followed when developing driver models:

- Every external port of the circuit should have a corresponding driver model port of the same name; the driver model cannot have any extra ports.

- The input/output direction of the driver model port should be opposite to that of the external circuit port, *e.g.* an INWARD driver would be paired with an OUTWARD port.

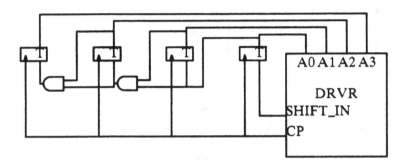

Figure 5-2. Four-bit binary counter

```
COMPTYPE drvr;
(* driver model for 4-bit counter circuit *)
  INWARD a0,a1,a2,a3:boolnet;
  OUTWARD cp,shift_in:boolnet;

  (* write out a net value to terminal *)
  PROCEDURE outvalue(val:boolean);
    BEGIN
    IF val THEN writeln('true')
    ELSE writeln('false');
    END;

  SUBPROCESS
    (* clock the counter every 20 nanoseconds *)
    stimulus: TRANSMIT NOT cp TO cp DELAY 10;

    (* watch for any changes in counter output *)
    monitor: UPON true CHECK a0,a1,a2,a3 DO
      BEGIN
      writeln('** Circuit Monitor at time ',TIME,' **');
      write('  A0 value: '); outval(a0);
      write('  A1 value: '); outval(a1);
      write('  A2 value: '); outval(a2);
      write('  A3 value: '); outval(a3);
      END;

  BEGIN
  ASSIGN true TO shift_in;
  END;
```

Figure 5-3. Driver Model for a Counter

Figure 5-1 shows a simple logic circuit. Notice that additional logic is included which connects a drvr component to the external connections of the circuit. In this case a driver model that monitors the output of the circuit is shown in Figure 5-3. In this model, the stimulus subprocess is used to drive the circuit, and a clock pulse with a period of 20 nanoseconds is generated. The monitor subprocess continually watches the outputs of the circuit, and whenever any of them changes value a message is written to the terminal.

The power of a driver model is derived from the expressive capability of the Pascal language. Any environment that can be described using Pascal code can be used to drive a design. For example, a customized assembly language parser can be included in a driver model to drive a circuit from a file containing assembly or microcode instructions. In addition, the output from the circuit can be written to a file with a format specific for a particular tester. The driver model allows the ADLIB user to create customized simulation environments around a particular circuit, and to increase the effectiveness of the simulations being performed.

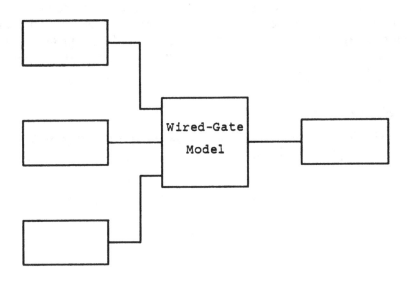

Figure 5-4. Wired-gate Modeling

5.3 Wired-Gate and Register Models

In ADLIB, all wired-gate and bus processing must be handled explicitly with wired-gate models. Figure 5-3 shows the approach taken to model a wired-gate or a bus. For every signal that has more than one source, a special model is inserted* to give the signal the special characteristics it requires. This approach is highly flexible, for the following reasons:

---

*   Insertion is automatic in the HELIX system.

- More than one model can be used, allowing special cases of wired gates or busses to be handled differently at different points in a circuit. For example, a circuit may use two different technologies (such as I^2L and TTL) which have different rules for wired gates. The ADLIB language can handle this situation with two different models.

- New technologies do not require a reworking of the internal simulator algorithm. Whenever a technology with new rules for handling busses or wired gates emerges, all that is required is to write a new wired-gate model which performs the correct operations.

- The user can customize the wired-gate processing to perform specialized error checking. For example, it may be desirable to report an error message whenever a bus has more than one simultaneous driving source. By modifying the ADLIB wired-gate model, it is possible to output an error message to the user which flags such a situation.

5.4 Wired-Gate Insertion

The following TTL example will illustrate the wired-gate processing. It is based on the simple TTL circuit shown below:

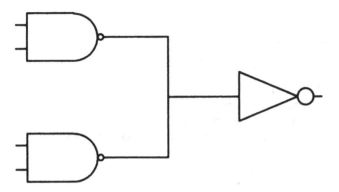

Figure 5-5. Wired-Gate Circuit Before Processing

The next Figure 5-shows the same circuit after a wired gate has been inserted.

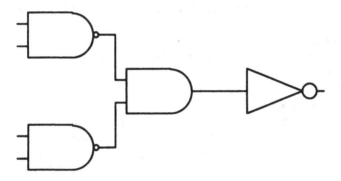

Figure 5-6. Wired-gate Circuit After Processing

```
NETTYPE
  netlogic=(unk,lo,hi,z);
COMPTYPE buscheck;
  INWARD a,b,c,d,e,f:netlogic;
  OUTWARD y:netlogic;
  VAR useval:netlogic; sources:integer;
  PROCEDURE countsrc(av,bv,cv,dv,ev,fv:netlogic;
                     var val:netlogic;var cnt:integer);
    BEGIN
    cnt:=0;
    val:=z;
    IF av<>z THEN BEGIN cnt:=cnt+1; val:=a; END;
    IF bv<>z THEN BEGIN cnt:=cnt+1; val:=b; END;
    IF cv<>z THEN BEGIN cnt:=cnt+1; val:=c; END;
    IF dv<>z THEN BEGIN cnt:=cnt+1; val:=d; END;
    IF ev<>z THEN BEGIN cnt:=cnt+1; val:=e; END;
    IF fv<>z THEN BEGIN cnt:=cnt+1; val:=f; END;
    END;
  BEGIN
  a:=z; b:=z; c:=z; d:=z; e:=z; f:=z;
  SENSITIZE(a,b,c,d,e,f);
  REPEAT
    DETACH;
     (* pickup count of z values *)
    countsrc(a,b,c,d,e,f,useval,sources);
     (* more than one driving net...return unknown *)
    IF sources>1 THEN ASSIGN unk TO y DELAY 0
     (* all high impedance...return high impedance *)
    ELSE IF sources=0 THEN ASSIGN z TO y DELAY 0
     (* single source...return value *)
    ELSE ASSIGN useval TO y DELAY 0;
  UNTIL false;
  END;
```

Figure 5-7. Sample Wired-Gate Model

A Wired-Gate Example

By supplying a model which performs the behavior associated with a bus or a wired-gate, the ADLIB language can handle a wide variety of special situations. For example, Figure 5-7 shows a typical model for a wired gate. This model performs the following functions:

- It watches for changes in the input ports.

- When all inputs are at high impedance, it assigns high impedance to the output.

- When exactly one input is driving, it propagates the input signal to the output.

- When more than one input is driving, it assigns an unknown value to output.

During initialization the model assigns a high impedance value to all ports. This approach insures that the model will operate correctly even if several input ports are unconnected, and never receive a new value from a source net. The countsrc routine is used to count the number of driving sources whenever the model is activated.

```
NETTYPE
  netlogic=(unk,lo,hi,z);
COMPTYPE res;
  INWARD inval:netlogic;
  BOTHWAYS outval:netlogic;
  BEGIN
  SENSITIZE(outval);
  REPEAT
    DETACH;
    IF outval=z THEN ASSIGN inval TO outval;
  UNTIL false;
  END;
```

Figure 5-8. Typical Resistor Model

Figure 5-8 illustrates a typical resistor model. In this case, the resistor propagates the inval value to its output port only if the output port value becomes high impedance.

The following rules apply to the wired-gate model and to the pullup resistor model:

- The wired-gate model must have only one output port.

- The wired-gate model must have at least as many input ports as the worst case multiple-source situation requires.

- Unconnected model ports should be handled correctly by the model (see later comments).

- The resistor model should passively pullup or pulldown the wired-gate output. In other words, only when the output value goes to high impedance should the resistor reassign a low or high value to the net (depending on whether the resistor is pullup or pulldown).

- The output port of the resistor should be defined in the model as BOTHWAYS.

The wired-gate model shown in Figure 5-7 and the resistor model shown in Figure 5-8 will operate correctly for most wired-gate and bus configurations. Modifying the models to perform special functions, such as reporting an error to the user when more than one source is driving the bus, is straightforward.

5.5 States and Memories

There are two ways to model memory in an ADLIB model:

1. The port values can be driven and later read.

2. Local variables can be declared inside a component type to store data.

A typical use of the first approach might be in a toggle flip-flop, as shown below:

```
NETTYPE
  boolnet=boolean;
COMPTYPE toggle;
  INWARD ck:boolnet;
  OUTWARD q:boolnet;
  SUBPROCESS
    clkdff: UPON ck CHECK ck DO
      ASSIGN NOT q TO q DELAY 10;
  BEGIN
  END;
```

Figure 5-9. Port Value Memory Example

In this case the output toggles on the rising edge of the clock. The q port is used to retain the previous toggle value and in this way saves the toggle state. Because the assign statement does not change the value of a port immediately, the q port can be used throughout the model when references to the current state of the flip-flop are required. Similarly, since the check clause guarantees that the clock has just changed value, if the clock is currently true the previous value of the clock must have been false, so the expression UPON ck causes the subprocess to be activated on the leading edge of the clock.

```
NETTYPE
  datapath=0..15;
  rammode=(readmem,writemem);

COMPTYPE ram;
  INWARD address:datapath;
          mode:rammode;
          datain:datapath;
  OUTWARD dataout:datapath;
  VAR mem:array[0..15]of datapath;
      i:integer;
  SUBPROCESS
    readram: UPON (mode=readmem) CHECK mode,address DO
      ASSIGN mem[address] TO dataout DELAY 30;
    writeram: UPON (mode=writemem)
        CHECK mode,address,datain DO
          mem[address]:=datain;
  BEGIN
  IF utlinit THEN
    FOR i:=0 TO 15 DO mem[i]:=0;
  END;
```

Figure 5-10. Local Variable Memory

On the other hand, local variables in a model are often the only convenient way of modeling the state of a complex hardware block. For example, the use of the local variable is essential to model random access memory (RAM). In Figure 5-10, the variable mem retains the random access memory values. An important differences between ADLIB and Pascal is found in the handling of local variables. These variables, such as mem, retain their values from one execution of an ADLIB model to the next execution. This contrasts with the way a Pascal procedure operates, since local variables lose their values from one call to the next call. This particular feature is essential to the utilization of local variables in ADLIB as state holders, and is one of the reasons that ADLIB is described as a *process-oriented* language.

In ROM and PLA models, it is often convenient to load the memory arrays from a file. The following ROM model illustrates the use of input/output statements to initialize a ROM model memory.

```
PACKAGE rompack;
  VAR infile:TEXT;
  BEGIN
  END.
```

Figure 5-11. ROM Package Definition

```
MODULE ramdemo;
  USE utl_pack,rompack;
  NETTYPE datapath=integer;
          address=0..15;

  COMPTYPE rom;
    INWARD addr:address;
    OUTWARD data:datapath;
    VAR memory:array[address]of datapath;
        i:integer;

    SUBPROCESS
      watch_address: UPON true CHECK addr DO
                  ASSIGN memory[addr] TO data DELAY 35;

    BEGIN
    (* initialize the ROM memory *)
    IF utlinit AND NOT utlrestart THEN
      BEGIN
      utlopen(infile,'ROMDEMO.ROM',utlold);

      (* read in the ROM values *)
      FOR i:=0 TO 15 DO readln(infile,memory[i]);

      (* close the ROM memory file *)
      utlclose(infile);
      END;
    END;
```

Figure 5-12. ROM Model

Figure 5-11 shows a simple package used by the ROM model, and Figure 5-12 shows the model for the ROM itself. The main body of the model opens a file named "ROMDEMO.ROM" and reads in 16 memory values. The watch_address subprocess continually watches for changes in the address input, and updates the data output appropriately. Figure 5-13 shows a sample data file used to initialize the ROM model. The format of the file is arbitrary, so by enhancing the ROM model it is possible to read more complex values from such a file, including such things as mnemonic micro-code instructions.

```
      12
      34
      19
      12
      10
       1
       0                .
      12
     100
     200
     300
     400
     500
     600
     700
     800
```

Figure 5-13. ROM Initialization Data

5.6 Synchronous and Asynchronous Processing

As will be discussed in Chapter 6, the simulator has two scheduling phases:

1. Update all net values.

2. Activate all components and subprocesses.

The characteristics of the scheduling mechanism can be used to advantage when writing ADLIB models. The model writer can assume that all subprocesses will have access to the same value of a given net at the same time. In addition, all processes which are synchronized to a given signal will always detect the change of the signal and the associated new value at the same time. The next three small examples show how this knowledge can be used.

```
posedge: UPON clk CHECK clk DO
   BEGIN
   ...
   END;
```

Figure 5-14. Positive Edge Triggering Subprocess

Figure 5-14 illustrates the modeling of a positive-edge-triggered
subprocess. The guard is tested whenever the `clk` port changes value, but
the subprocess body is activated only when the new clock value is `true`,
and consequently only on the leading edge of a clock pulse.

```
negedge: UPON NOT clk CHECK clk DO
   BEGIN
   ...
   END;
```

Figure 5-15. Negative-Edge-Triggering Subprocess

Figure 5-15 illustrates a negative-edge-triggered subprocess. The guard
`NOT clk` is tested whenever the clock changes value, but allows
subprocess to be activated only when the clock is `false`, and
consequently only on the trailing edge of a clock pulse.

```
COMPTYPE dlatch;
   INWARD enable:boolnet;
            d:boolnet;
   OUTWARD q:boolnet;

   SUBPROCESS
      watchinput: UPON enable CHECK enable,d DO
                   ASSIGN d TO q DELAY 10;

   BEGIN
   END;
```

Figure 5-16. Asynchronous Triggering

Finally, when modeling asynchronous processes, the subprocess must be
sensitive to all possible inputs. For example, Figure 5-16 shows a D latch.
In this model, the output continually follows the input whenever a change
is seen while `enable` is `true`, but will latch onto the last value seen and
hold it while `enable` is `false`. The enabled mode of operation
demonstrates asynchronous operation, since the `enable` and d inputs can
change at any time and the model will respond accordingly.

5.7 Use of Concurrent Processes

The appropriate use of subprocesses is important both from an efficiency
standpoint as well as a model readability standpoint in ADLIB. The
following are good rules of thumb for deciding on the number of
subprocesses to use in writing models:

- Each synchronous operation can typically be modeled using a single subprocess

- Each asynchronous operation (such as a clear or preset input) will typically require a separate subprocess

- Occasionally the most convenient way of modeling multi-mode devices (such as read mode/write mode), is to create one subprocess per mode.

```
NETTYPE boolnet=boolean;

COMPTYPE jkff;
  INWARD j,k,pr,clr,ck:boolnet;
  OUTWARD q:boolnet;

SUBPROCESS
  preset: UPON (NOT pr) CHECK pr DO
               ASSIGN true TO q DELAY 10;

  clear: UPON (NOT clr) CHECK clk DO
               ASSIGN false TO q DELAY 10;

  jk: UPON (NOT ck)AND(pr)AND(clr) CHECK ck DO
          BEGIN
          IF j AND k THEN ASSIGN (NOT q) TO q DELAY 10
          ELSE IF (NOT j)AND k THEN
            ASSIGN false TO q DELAY 10
          ELSE IF j AND (NOT k) THEN
            ASSIGN true TO q DELAY 10;
          END;

  BEGIN
  END;
```

Figure 5-17. Asynchronous Processes

Figure 5-17 shows a JK flip-flop model which illustrates these concepts. The flip-flop has two asynchronous inputs, pr and clr which are both "active low." In addition, the JK function is synchronous to the negative edge of the clock ck. In this case, the rule-of-thumb indicates that three subprocesses would be convenient, and, in fact, the above model is reasonably efficient and readable.

A more advanced use of subprocesses in ADLIB involves the use of internal nets. In the previous chapter a network communication algorithm was described. As part of the interrupt processing of this algorithm, an internal net and an associated subprocess were used. In general, a subprocess is used for each concurrent activity that a hardware block performs. In addition, special purpose processes are used to handle interrupts, input/output functions and error checking.

5.8 Pointers in Net Values

A net type can be declared to be a record type containing one or more pointer (dynamically allocated) types, as shown in the following example:

```
TYPE
   netvalueinfo=RECORD
      load:integer;
      usage:integer;
      END;
  NETTYPE
    rectype=RECORD
       f1:integer;
       f2:^netvalueinfo;
       END;
```

Figure 5-18. Net Type with Pointer

Figure 5-18 shows a net type rectype that includes a field f2 that is a pointer to the record type netvalueinfo. The use of pointers in net types can be thought of as attaching a *net value attribute* to nets of this type. In the example cited here, the load and usage values represent additional information about the net. The models can access this information, but the simulator will ignore it. More importantly, the data structure can be dynamic (such as a linked list of records) and can grow as the information associated with the net grows. Later examples which discuss timing verification and fault simulation in ADLIB will illustrate this.

```
COMPTYPE useptr;
  BOTHWAYS ptrpin:rectype;
   ...
  BEGIN
  IF utlinit THEN
    BEGIN
    IF ptrpin.f2<>nil THEN
      BEGIN
      new(ptrpin.f2);
      ptrpin.f2^.load:=0;
      ptrpin.f2^.usage:=0;
      END;
    END;
  END;
```

Figure 5-19. Pointer Initialization

Initialization of pointer values must be done by the models connected to
these types of nets, and Figure 5-19 illustrates the correct way to do this.
Notice the test for a non-nil value of the f2 field. The simulator will
initialize all pointer fields to nil before the start of simulation, but since
the net is connected to more than one component, the model cannot
predict which model will allocate the pointer reference first.

Assignment of net types that use pointers should be performed carefully,
so that the pointer value will not be lost. For example, the following
code:

```
INWARD ptrpin:rectype;
 VAR temp:rectype;

temp.f1:=ptrpin.f1;
temp.f2:=ptrpin.f2;
 .
 .
 .

ASSIGN temp TO ptrpin;
```
Figure 5-20. Assignment of Pointer Net Types

will maintain the pointer value of the port, but the model:

```
INWARD ptrpin:rectype;
 VAR temp:rectype;

 temp.f1:=nil;
 temp.f2:=ptrpin.f2;
  .
  .
  .

 ASSIGN temp TO ptrpin;
```

destroys the pointer value, and the referenced memory will be lost.

5.9 Model Initialization

Typically, the main body of a model is used for model initialization. The model writer should be aware that:

- the main model body is called twice, once at the beginning of the initialization phase, and once at the beginning of the simulation phase.

- the order in which each component in a circuit is initialized is arbitrary, and is not under control of the user.

```
NETTYPE
   datapath=0..15;

COMPTYPE reg;
   ...
   OUTWARD dataout:datapath;
   ...
BEGIN
IF utlinit AND NOT utlrestart THEN
   ASSIGN 0 TO dataout;
END;
```

Figure 5-21. Model Initialization

Figure 5-21 illustrates a typical initialization body for a model. The utlinit routine returns true during the initialization phase of the simulation, and the utlrestart routine returns true if the simulator is restarting. The assign statement schedules an update to the output net with zero delay; this approach may cause other components which are

```
   ...
BEGIN
IF utlinit AND NOT utlrestart THEN dataout:=0;
END;
```

Figure 5-22. Immediate Net Value Assignment

connected to this net to be activated. If it is undesirable to cause other activity in the circuit, an immediate net value assignment can be used as shown in Figure 5-22. The output net will get a value of zero during initialization, but no components connected to the net will be activated. The immediate net value assignment changes the net value without use of the event queue, and consequently, no activities are scheduled.

Local variables used in models should always be initialized since The simulator will not initialize them.

```
NETTYPE
   datapath=0..15;

COMPTYPE ram;
   ...
   VAR mem:array[0..15]of datapath;
       i:integer;
   ...
   BEGIN
   IF utlinit AND restart THEN
     FOR i:=0 TO 15 DO mem[i]:=0;
   END;
```

Figure 5-23. Initialization of Local Variables

Figure 5-23 illustrates the initialization of the local variable mem which is used during simulation to save the model memory values. The initialization code in the main body of this model initializes all of memory to zero. If this statement had not been executed, the memory would contain random values, many of which would not be in the range 0..15. Consequently, values read from the memory during initialization would be invalid, and the simulation might fail with a run-time fault.

When writing models, the user should be concerned about the startup characteristics of the models and should be aware that the restart command from the simulator will reinitialize all values, unless tested for with the function utlrestart. The simulator will initialize various different net types in different ways. For example, integer net types will be initialized to zero.

Net Data Type	Initial Value
Boolean	False
Integer	0
Real	0.0
Enumerated	First element of enumeration list
Subrange	First element of subrange
Record	Each field handled separately
Array	Each element handled separately
Pointers	Not initialized
Char	ord(0)
Set	Empty Set

Figure 5-24. Simulator Initialization of Net Values

Figure 5-24 shows the defaults that the simulator uses when initializing various net types in ADLIB. In writing models, the model writer should be aware of the initial value of nets. Typically, in logic simulation, the unknown value should be the first value in the list of possible values for nets, so that all nets will be initialized to unknown prior to the start of simulation. In addition, the model writer should be concerned about unknown handling and initialization handling to ensure that a worst-case result is always output. For example, for an AND gate with the following inputs:

    input 1 = unknown
    input 2 = high

the output should be unknown since the value of input 1 can influence the output of the gate. If in doubt, the modeler should choose a behavior in which errors will be most easily detected in simulation output.

5.10 Checking Timing Constraints

The ADLIB language has been used to perform timing verification in recent work at Silvar-Lisco [COE84]. This section will summarize the approach and the results of this effort.

Three major approaches have been used to verify the timing constraints of digital systems. Logic simulation [KUS1976], [SZY72], worst-case path analysis [WOL78], and timing verification [MCW80][AGR82]. Logic simulation requires either a complete design or the generation of input test vectors that would drive the undefined signals. This makes it hard to use early in the design process, and can lead to long simulation run times when large sets of test vectors are used. Worst-case path analysis examines all paths through the combinational logic between registers or latches, searching for the longest and shortest paths. This approach fails with sequential circuits, and also tends to be expensive. For these reasons, the timing verification algorithm and was the approach taken here.

5.11 Timing Verification Algorithm

The basic technique in the timing verification algorithm is to identify when signals are changing and when they are stable, and to measure this time relative to the basic clock period. This information is then used to check for set-up, hold, minimum pulse-width, and bus contention errors. For instance, to determine if there is a setup violation, it is sufficient to see if the input signal is stable for the duration of the minimum setup requirement before it is clocked.

Signals are allowed to have one of the following values: *unknown, changing, stable, zero, one,* or high impedance. Typically, however, only *changing* and *stable* are used. The *changing* value represents a signal that can be changing from one to zero, or from zero to one; the *stable* value represents a signal that is constant, but whose actual value (zero or one) is not known. Occasionally, the timing of a circuit will depend on the values of signals, in which case actual values are used (unknown, zero, one).

The value of a signal over the clock period is represented by a linked list, each node of which specifies a value and the duration of that value. The sum of the durations of all the nodes in the list must equal the period (cycle time) of the circuit.

As illustrated in Chapter 4, the timing parameters such as minimum and maximum propagation delays, setup and hold times, clock-pulse width constraints, and hazard times, for specific components in a design are entered as attributes in the structure specification, and passed to the behavior models through their parameters. Default values for these parameters, however, may be specified directly in the ADLIB models.

Timing models (or *timing primitives*) are descriptions of combinational gates and synchronous parts such as flip-flops and registers, writen in ADLIB with built-in timing-constraint checking. Unlike the SCALD timing verification implementation [MCW80], timing-constraint checking is built into the description of the model, along with its logical behavior. This means that timing errors can be checked while signals are being propagated, eliminating the need to check for timing errors after the signals have stabilized. This is a particular advantage in circuits with few feedback paths.

The type declarations for the timing models are shown below:

```
tvalptr = ^tvalrec;
   tvalrec = RECORD;
           val:(u,0,1,S,C,HIZ);(* value *)
           delta:integer;(* duration *)
           link:tvalptr; (* link to next val *)
           END;
   timingstate = RECORD
                   changed:boolean;
                   state:tvalptr;
                   END;
   NETTYPE timingnet=timingstate;
```

Figure 5-25. Timing Model Type Definitions

These definitions are used in Figure 5-26, which shows the timing model for an inverter. The net type timingnet is of the type timingstate which has two fields, called changed and state. The state field references a linked list, each node of which represents a value and its duration. The changed field is a boolean flag which indicates a change in the timing list value of the signal. It is this ability to define nets having complex data types that permits the writing of timing models in ADLIB.

The model described in Figure 5-26 performs two functions:

1. It propagates the timing list.

2. It checks the timing constraints.

The default timing parameters, cdmin and cdmax are the component delays. During initialization, the model sets the output to an unknown timing state and initializes the local variables. When a change is detected in the input the subprocess compop is activated. The procedure l_update_timing_state determines the output timing state from the input using a table-lookup operation. (All combinational elements are defined as tables.) The l_delay_timing_state procedure incorporates the component delay in the timing-state, by introducing an unstable

```
COMPTYPE inverter( cdmin,cdmax,haztim: integer);
   (* define default delays and hazard time *)
  DEFAULT cdmin=3; cdmax=7 ;haztim=4;
     (* declare the pins on model and
        ***      associated data types *)
  INWARD a : timingnet;
  OUTWARD z : timingnet;
     (* declare local variables *)
  VAR newval, delayval : l_timingstate;
      out_num:integer;
      out_name:utl_string20;
SUBPROCESS
 (* declare main process *)
 compop: UPON true CHECK a DO
  BEGIN
    (* determine the output timing state *)
    l_update_timing_state(invtype,a,a,newval);
    (* add delay to the output timing state *)
    l_delay_timing_state(newval,delayval,cdmin,cdmax);
    (* check hazards at the output *)
    check_hazard(delayval.state,out_name,haztim);
    (* assign the new timing state to output *)
    ASSIGN delayval TO z;
    END;
BEGIN
(* initialize the model *)
IF utlinit THEN
   BEGIN
   ASSIGN l_start_value TO z;
   newval.changed:=true;
   newval.state:=nil;
   utlidnet(z,out_name,out_num);
   END;
END;
```

Figure 5-26. ADLIB Timing Model for Inverter

region of period equal to the difference between cdmax (max delay) and cdmin (min delay). The procedure check_hazard checks if any '0' or '1' hazards appear in the output timing state, and flags an error if necessary. This provides a warning to the user if any glitches, hazards, or events faster then the frequency response of the component occur at the input. In this example, the parameter haz_tim represents the *inertial delay* of the component, which specifies the minimum input pulse width allowed.

An additional advantage of doing timing-constraint checking within the component descriptions is that signals requiring case analysis need not be specified explicitly by the user, since the correct action can be determined directly by the nature of the component. For example, a multiplexer's selection input could determine which of its input signals should be routed to its output. A stable, but logically unknown, value on the selection input could be evaluated for both 'zero' and 'one' cases, and could generate two events at the output. This would permit simultaneous checking of both paths through the multiplexer.

Currently, a timing verification library has been written supporting a variety of SSI and MSI parts, including AND, OR, and INVERT gates, D-latches, JK flip-flops, and RS flip-flops. The nature of the system is such that it is easy to build upon. Typically, an SSI model requires 1 to 2 staff-days of effort to code and debug, and an MSI part requires 1 to 5 staff-days. Since a high-level hardware description language is used, designers can write their own timing models to meet their needs; modeling does not require an expert intimately familiar with the internals of the simulator. As more models are written, they can be collected into large libraries. In previous approaches this would have been prohibitively expensive and time consuming.

5.12 Timing Assertion Input

Simulation stimulus is made via an assertion file that specifies the times when input signals are stable or changing, the clock period for which the timing verification is performed, and the interface signal specification that is to be used.

```
ASSERT signalname <time1> <time2> <val>..
REPEAT clockname <start> <stop> <offset> <val>..
where:

        <start>   - startime (default=zero)
        <stop>    - stoptime ( default=cycletime )
        <offset>  - duration of val
        <val>     - value (0,1,S,C,U,Z)
        <time1>   - start of duration
        <time2>   - end  of duration
```

Figure 5-27. Signal Assertions

```
CYCLETIME 300;
  ASSERT VDD 0 300 s;
  ASSERT set 0 10 one 10 300 zero;
  REPEAT clk 0 300 80 one 80 zero;
```
Figure 5-28. Typical Assertion Input

The format of the control, data signal assertions and clock assertions is shown in Figure 5-27. Typically, only clocks that are much faster than the base clock period need be specified by a REPEAT assertion. All unspecified periods in an assertion are assumed to be changing. One typical assertion input used for the modulo-5 counter circuit is described in Figure 5-28. The first statement specifies the cycle time for verification, and the next three statements assert the primary inputs (vdd and set) and the clock (clk).

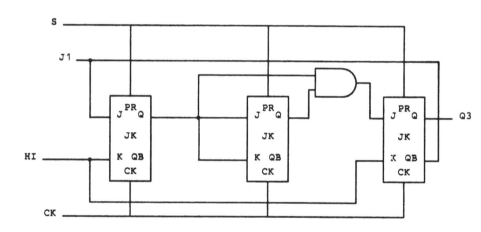

Figure 5-29. A Modulo-5 Counter

5.13 Performance Comparison

The speed of a timing verifier can be critical to its utility − if it is too slow, it can be difficult or impossible to use on the large circuits that really need it. To evaluate the effectiveness of ADLIB for timing verification, its speed has been compared with that of the SCALD [MCW80] timing verifier. Figure 5-29 shows the example used for comparison. It is a modulo-5 counter implemented using JK flip-flops. To perform timing verification in SCALD, the JK flip-flops were expanded down to the level of the RS register primitive in SCALD, resulting in an increased component and net count. In contrast, the circuit was verified as JK flip-flops in ADLIB.

	SCALD	ADLIB/HELIX
Setup CPU time(sec)	7.13	2.81
Simulation CPU time (sec)	2.77	0.53
Events/CPU second	12	63

Figure 5-30. Speed Comparison of SCALD vs ADLIB/HELIX

Figure 5-30 shows the results of timing verification runs using the example shown in Figure 5-29, as performed on a VAX 11/750 system. The simulation CPU time indicates that ADLIB/HELIX timing verification is about five times faster than SCALD timing verification. A similar improvement is seen in the number of events processed per CPU second, a factor closely related to the simulation time. This improvement in speed is not the upper limit and may actually be larger for more complex MSI parts.

5.14 Fault Simulation

Fault simulation is used to predict the effectiveness of a given set of test vectors in detecting circuit failures. The most common model used for circuit failures is the stuck-at fault, in which a circuit failure is assumed to appear as a frozen net, with a value of 1 or 0. The ADLIB language has been utilized in fault simulation research at Silvar-Lisco based on this methodology.

The approach taken is summarized here:

- Start with a given test vector, and generate a fault list for the nets of each gate. To provoke the faults, assign the complement of the correct value.

- Evaluate the components connected to the inputs, and propagate the correct values and fault lists to the outputs of the components. The input and output fault lists are merged, so that outputs which can cause faulty output are propagated, but those which do not generate faults are not propagated.

- These lists are allowed to propagate through the circuit in an event-driven fashion.

```
NETTYPE
  flt=(s_a_0,s_a_1);
  en_val=(u,zero,one);
  qptr=^q;
  q=RECORD
    node_num:integer;
    type_fault:flt;
    eq_flt:qptr;
    nxt:qptr;
    END;
  intnet_val=RECORD
    val:en_val;
    lptr:qptr;
    END;
```

Figure 5-31. Fault Simulation Net Type Definitions

For each net in the circuit, a list of faults is maintained that records all faults capable of causing the output value to be different from its correct value. All of these functions are performed in ADLIB models associated with individual components in a circuit, not with the simulator itself. Figure 5-31 shows the net types associated with the fault simulation models. The net type used to connect components is called intnet_val, and it is a record type composed of a field with the correct value val, and another field lptr which contains a list of faults that would make the current node faulty.

```
COMPTYPE and2;
INWARD a,b:intnet;
OUTWARD y:intnet;
VAR
 temp:en_val;
 num:integer;
 namea,nameb,nameout:utl_string20;
 tmpnet:intnet_val;
SUBPROCESS
   fltprop: UPON true CHECK a,b DO
      BEGIN
      IF (a.val<>u) AND (b.val<>u) THEN
         BEGIN
         (* create input fault lists *)
         flt_crlist(a,namea) ;
         flt_crlist(b,nameb) ;
         (* evaluate correct circuit value at output *)
         tmpnet.val:=logtab[andtyp,a.val,b.val];
         flt_crlist(tmpnet,nameout);
         IF (a.val=one) AND (b.val=one) THEN
            BEGIN
            merge_lists(a,tmpnet); merge_lists(b,tmpnet);
            END
         ELSE IF (a.val=one) AND (b.val=zero)
            THEN merge_lists(b,tmpnet)
         ELSE IF (a.val=zero) AND (b.val=one)
            THEN merge_lists(a,tmpnet);
         ASSIGN tmpnet TO y DELAY 1;
         WAITFOR true DELAY 1;
         print_out(a,nameA);
         print_out(b,nameB);
         print_out(y,nameout);
         END;
      END;
BEGIN
IF utlinit THEN
(* initialize local variable *)
    BEGIN
    tmpnet.lptr:=nil;
    tmpnet.val:=u;
    END;
(* pickup net names and net numbers *)
utlidnet(a,namea,num);
utlidnet(b,nameb,num);
utlidnet(y,nameout,num);
END;
```

Figure 5-32. AND2 Fault Model

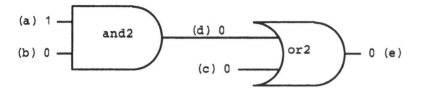

Figure 5-33. Fault Simulation Example - Copy 1

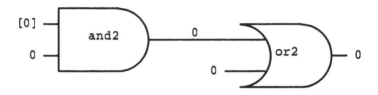

Figure 5-34. Fault List Copy 2

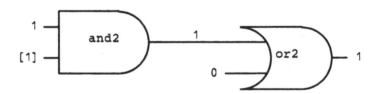

Figure 5-35. Fault List Copy 3

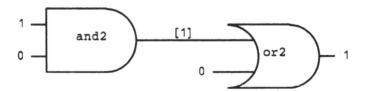

Figure 5-36. Fault List Copy 4

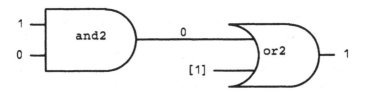

Figure 5-37. Fault List Copy 5

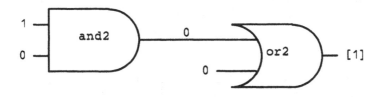

Figure 5-38. Fault List Copy 6

Figure 5-32 shows the fault simulation model for a two input AND gate. To illustrate how this gate works, consider the case shown in Figure 5-33. The six copies considered for concurrent fault simulation, including the correct values, are shown in Figures 33 through 37. The following summarizes the operation of the fault simulation algorithm:

Fault Generation: In ADLIB the evaluation is performed first for and2. Three of its nodes can be stuck-at-fault values, namely a(s-a-0), b(s-a-1) and d(s-a-1).

Fault dropping and merging: Copies 3 and 4 force a 1 on the node d, whereas copy 2 forces node d to zero (the correct value). Hence b(s-a-1) and d(s-a-1) are stored in the fault list for node d and a(s-a-0) is dropped.

Fault assignment and propagation through the circuit: This newly created list is assigned to node d causing component or2 to be activated. Copies 4, 5, and 6 are created corresponding to d(s-a-1), c(s-a-1) and e(s-a-1). In each of these situations node e is forced to a faulty value. However a previously created fault list for node d exists, and that forces it to a fault value of 1. If ever d(s-a-1) is detected at a later node, then all the faults leading to a value of 1 on node d will also be detected. Hence the fault list at d, c(s-a-1) and e(s-a-1) are stored in the fault list of e. This merger of fault lists at d, c and e leads to the following list of detectable faults at e: b(s-a-1), d(s-a-1), c(s-a-1) and e(s-a-1).

For a given test vector, all possible faults are evaluated for all components. In this process, each component is evaluated for the correct circuit and all faulty circuits while simultaneously carrying out fault-dropping and fault-collapsing. Running several test vectors through the circuit in a single simulation automatically ensures coverage of a large number of faults, without reevaluation of those nodes whose value remains the same for different test vectors. (This is a natural consequence of the event-driven nature of ADLIB.) Fault collapsing is achieved by keeping a list of equivalent faults at each node and not evaluating the corresponding faults. This is done in procedure flt_crlist in the example shown above.

5.15 Symbolic Simulation

The ADLIB language has also been used to perform *symbolic* simulation. James King developed symbolic simulation [KIN75] [KIN76] for the testing of computer programs. Warren Cory [COR81] demonstrated that ADLIB can be used to simulate electronic hardware symbolically.

At any time during conventional simulation, the nets in a circuit have known constant values (*e.g.* 1 or 0). In such a simulation, every component in a circuit will produce constant outputs given constant inputs. On the other hand, in symbolic simulation, the nets in a circuit have *symbolic* values (*e.g.* "x" or "2*z"). A symbolic value represents a constant but unknown value. In symbolic simulation, components that take symbolic inputs will return symbolic outputs. The advantage of symbolic simulation is that all possible values for nets will be tested simultaneously, while during conventional simulation only one specific test case can be evaluated at a time.

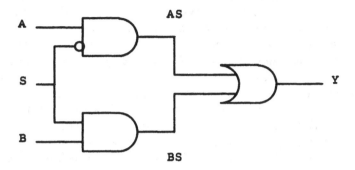

Figure 5-39. Simple Electronic Circuit

time	A	B	S	AS	BS	Y
0	F	F	F	F	F	F
10	T	F	F	F	F	F
11	T	F	F	T	F	F
12	T	F	F	T	F	T
20	T	F	T	T	F	T
21	T	F	T	F	F	T
22	T	F	T	F	F	F
30	T	T	T	F	F	F
31	T	T	T	F	T	F
32	T	T	T	F	T	T

Figure 5-40. Sequence of Inputs and Outputs

```
time  A    B    S        AS              BS                        Y
0    %X   %Y   %Z   ( %X & -%Z) (%Y &   %Z) ( %X & -%Z)^(%Y &   %Z)
10   -%X  %Y   %Z   (-%X & -%Z) (%Y &   %Z) (-%X & -%Z)^(%Y &   %Z)
20   -%X  %Y   -%Z  (-%X &  %Z) (%Y & -%Z) (-%X &  %Z)^(%Y & -%Z)
                         (etc...)
```

Figure 5-41. Equivalent Symbolic Simulation

Figure 5-39 shows a simple gate-level circuit, and Figure 5-40 shows a sequence of inputs and outputs for the circuit. This sequence tests one possible set of inputs. For this example, the propagation time through the gates is one time unit. In contrast, Figure 5-41 shows the equivalent symbolic simulation of this same circuit. The inputs to the simulation at time zero are the symbolic variables %X, %Y and %Z. At time 10, the value of the first input changes to the inverse of %X, which is written as -%X, and the effect on the output is immediate. (In this case, the symbolic simulation was designed to respond with zero time delay, although that is not required.) For any input, the circuit gives an output that is a combinational function of the symbolic values of the input. Although this is a trivial example, it does demonstrate the power of the symbolic simulation in analyzing a circuit. A more interesting case might involve the introduction of feedback loops and of memory in a circuit. In cases such as these symbolic simulation can perform an analysis of a circuit which might otherwise be too difficult for other formal analysis approaches to handle.

```
CONST symblo=0;
      symbhi=30;
TYPE nibble=0..15;
     alph=packed array[1..5]of char;
     ops=(boole,integ,xvar,xord,xodd,xnot,xand,xor,
          times,xdiv, xmod,plus,negate,minus,equal,
          ntequal,greater,less, grequal,lsequal);
     expression=RECORD
       operator:ops;
       CASE integer OF
         0: (left,right: ^expression);
         1: (name: alph);
         2: (b:boolean);
         3: (i:integer);
         END;
     eptr=^expression;
NETTYPE
     symbolicnet=eptr;
VAR
     fsh,symbtrue,symbfalse:eptr;
     symbconst:array[symblo..symphi]of eptr;
     index:integer;
     unaryops,binaryops,booleanops,integerops:SET OF ops;
```

Figure 5-42. Symbolic Simulation Type Definitions

```
BEGIN
fsh:=nil;
newe(symbtrue);
WITH symbtrue^ DO
  BEGIN
  symbtrue^.operator:=boole;
  symbtrue^.b:=true;
  END;
newe(symbfalse);
WITH symbfalse^ DO
  BEGIN
  symbfalse^.operator:=boole;
  symbfalse^.b:=false;
  END;
FOR index:=symblo TO symbhi DO
  BEGIN
  newe(symbconst[index]);
  symbconst[index]^.operator:=integ;
  symbconst[index]^.i:=index;
  END;
unaryops:=[xodd,xord,xnot,negate];
binaryops:=[xand,xor,times,xdiv,xmod,plus,minus,equal,
            ntequal,greater,less,grequal,lsequal];
booleanops:=[xodd,xnot,xand,xor,equal,ntequal,greater,
            less,grequal,lsequal];
integerops:=[xord,times,xdiv,xmod,plus,negate,minus];
END.
```

Figure 5-43. Global Initialization for Symbolic Simulation

```
COMPTYPE nor(pd:integer);
  DEFAULT pd=15;
  INWARD a,b:symbolicnet;
  OUTWARD y:symbolicnet;
  BEGIN
  ASSIGN nil TO y;
  WAITFOR DELAY 0;
  REPEAT
    assignsymb(ux(xnot, bx(xor, a, b)), y, pd);
    WAITFOR true CHECK a,b;
  UNTIL false;
  END;

COMPTYPE fourbitcntr;
  INWARD clk:symbolicnet;
  BOTHWAYS q:symbolicnet;
  SUBPROCESS
    symcheck: UPON NOT econst(clk) CHECK clk DO
      writeln('*4-bit Counter Not Constant At ',TIME);
  BEGIN
  ASSIGN nil TO q;
  WAITFOR DELAY 0;
  WHILE true DO
    BEGIN
    WAITFOR etru(clk) CHECK clk;
    assignsymb(bx(xmod, bx(plus, q, ic(1)), ic(16)), q, 0);
    END;
  END;
```

Figure 5-44. Symbolic Simulation Models

Figure 5-42 shows the ADLIB type definitions and global variable declarations for the symbolic simulation models. Figure 5-43 shows the initialization for the global variables used, and the support routines for symbolic simulation are found in the Appendix. Finally, Figure 5-44 shows two symbolic simulation models, nor and fourbitcntr. In the nor model, the inputs a and b are continuously monitored. Whenever either of them changes, the symbolic expression for the NOR gate is transmitted given the two input expressions. The fourbitcntr model waits for the clk input to become true, and increments the output by one, modulo 16.

Summary — Extended Uses for ADLIB

ADLIB's ability to support symbolic simulation rests on its facilities for declaring new net types, especially dynamic net types. This same capability was critical to both the timing verification and the fault simulation packages illustrated in this chapter.

CHAPTER 6

IMPLEMENTATION

6.1 Why Study an Implementation?

Sometimes understanding how a concept is implemented can help
illustrate the concept itself. That is the motivation behind this chapter,
which discusses a specific implementation of the ADLIB language and its
simulation environment. That implementation is the one done at
Stanford on the Hamburg Pascal compiler for the DEC-10.

6.2 Sequence of System Programs

The sequence of operations required to simulate an ADLIB program is as
follows: the ADLIB source is precompiled into a Pascal file and an
interface database file. The Pascal source code includes text that mirrors
the ADLIB source directly, plus code to support some of the simulation
setup and run-time operations. Since this support code is keyed to the
user-defined net types and the header (*i.e.* the externally visible part) of
each component type. it must be custom generated for each ADLIB
source file. The custom Pascal code is then compiled and linked with the
remainder of the run-time system, which is independent of the ADLIB
source. During simulation initialization the structure of the target system
is read in and simulation begins. Actual simulation operates under

---

*   There have been several other implementations of ADLIB over the years, some more
    sophisticated than others. The more sophisticated systems include interactive debugging
    and graphical interfaces; they have been omitted since they are host specific and beyond
    the scope of this book.

interactive control, with each event triggering and being triggered by a set of software "processes" which are multiprogrammed under the simulation control monitor. This sequence is illustrated below:

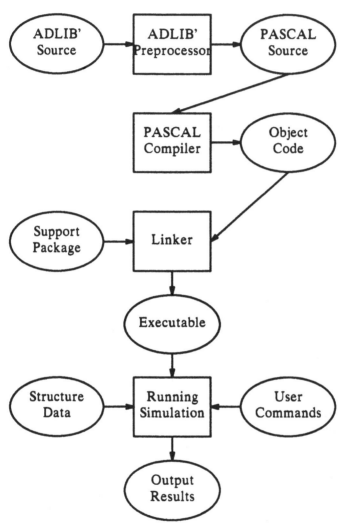

Figure 6-1. Structure of ADLIB Support

6.3 Preprocessing ADLIB into Pascal

The ADLIB precompiler is built around a one-pass, recursive-descent parser**. As it scans the ADLIB source file, one token at a time, it builds internal data structures and occasionally emits Pascal code and other information. In the process it does ADLIB syntax checking and type checking and partial Pascal type checking. This is accomplished by keeping a symbol table containing all the relations between net types and ordinary types. Experience has shown that it is important to do a fairly complete checkout here, since syntax errors caught during the subsequent Pascal compilation phase are difficult to trace back to the offending line in the original ADLIB source***.

In addition to the typical compiler data structures, the ADLIB preprocessor must keep a list of output token records for the body of each component type. This list is allocated dynamically and grows in proportion to the length of the longest component type body in the source file. The reason for needing this list is that some of the transformations from ADLIB to Pascal imply considerable latency. In particular, the multiprogramming aspects of ADLIB require that all the WAITFOR statements in a component type be identified before the code for the component type can be written out.

A Note About Pascal Typechecking and Name Conflicts

In the ADLIB simulation environment it is sometimes necessary to defeat the strong typing provided by Pascal. This is because the support environment must be able to store and copy pointers to any type of data structure that the user may declare. The support code uses "pointer to integer" type (integer_ptr) as a dummy type for this purpose, and coerces it back and forth to the "pointer to user-declared structure" type before dereferencing. This is always done with variant records. For example, the following code would be used to coerce a pointer to a string into a pointer to an integer,

---

** Recursive descent is a parsing technique wherein the structure of the parser models closely the productions in the language grammer. Thus the parser contains routines with names like parse_statement parse_comptype parse_argument_list, *etc*. This technique was chosen because of its simplicity and because it did not require an external parser generator, which might limit portability.

*** The original DEC-10 version did a general syntax verification and type check, HELIX does a complete one.

```
TYPE
integer_ptr = ^integer;

user_data = RECORD
     name : char array [1..10];
     END;

user_data_ptr = ^user_data;

dummy_type = RECORD
     CASE boolean OF
          true: (users: user_data_ptr);
          false: (fake: integer_ptr);
          END;
     END;

VAR
switch : dummy_type;

PROCEDURE put_num(p : integer_ptr);
     BEGIN
     dummy.fake := p; (*COERCION*)
     write(p^.user_data^.name);
     END;

BEGIN
switch.user_data := 'hello guy ';
put_num(switch.fake);
END;
```

Another problem comes with the conflict of user declared names and those needed by the run-time support environment. The approach taken is to have the names in the support environment always begin with the prefix "z9". This prefix is specifically forbidden in user supplied code. In addition, some of the support routines take names that are reserved words in ADLIB such as ASSIGN, since user supplied names can never conflict with these.

In some parts of the simulation environment, it is necessary to refer to objects by number which were unnumbered in the original source. For example: "the 5th net type," or "the 3rd net connection inside the 10th component, which is of the 2nd component type." Such access could be accomplished easily if all the references were keyed to arrays. However, to speed up and simplify the internal operations of components during simulation, the objects refer to each other by pointer. Thus a redundant array of pointers is necessary to provide access to data objects by number. In the case of ADLIB ports, this array of pointers is coerced onto the individual, symbolically named pointers, such as clockin, output, *etc.*, so that the support software can set up symbolic pointers. Likewise, numeric references to types are supported with CASE statements inside records; numeric references to component types is implemented with case statements keyed by the ordinal number of the item in the original source.

One additional feature of Pascal required by ADLIB is the ability to use labels and GOTO's to jump into and out of the structured program statements CASE, WHILE and REPEAT. Any Pascal compiler that puts severe restrictions on GOTO's would be difficult to work with.

6.4 Converting Component Types into Procedures

ADLIB component type definitions are completely rewritten into procedures (here called *component type procedures*) in the Pascal file output. The transformation is complex and requires that the complete component type definition be read in before the first line of the component type procedure be written out. The complexity stems from the process model of simulation; during simulation components must behave like independently executing processes under the control and coordination of the simulation monitor. This multiprogramming is achieved using labels, GOTO's and large record variables on the heap to retain the state of each component while it is not executing. This implementation technique does not give complete flexibility to the programmer since it does not provide the ability to stop a process inside a subroutine, as can be done in some other languages such as Concurrent Pascal, Ada, or Simula 67. However, it does seem to be adequate for many hardware description problems. It is also efficient, requiring just a few subroutine entries and CASE statements for each process switch.

User variable declarations, subroutines and executable code from a component type definition all go inside the component type procedure. The parameters to this procedure are (perhaps surprisingly) neither the ports nor the ADLIB parameters specified by the user for customizing each component instance. Rather, the only parameter to the component type procedure is a pointer to a Pascal record that represents the state of the component. This state carries within it pointers to the actual nets to which the component is connected, as well as the values of local component variables, the user specified parameters, and a field called delta which indicates where the component is to resume execution. In the Pascal code this record type definition is scoped inside the component type procedure. This means that components cannot possibly interfere, or even interact, with the internals of other components. Thus the implementation guarantees and takes advantage of the property of information hiding, which is basic to ADLIB.

While the component type definition is being scanned, all of its externally visible information including its name, its nets and their net type, and any customizing parameters such as latency, are extracted and written into a database file. This is the minimum information needed to link the ADLIB source with the external structure information.

In order for each component to access its local variables easily, the actual body of the component type definition, together with its subprocesses, goes inside a large WITH statement. The semantics of Pascal's WITH make access to the local storage almost completely transparent in the Pascal code produced. Thus the component type definition

```
COMPTYPE foo
    NET
            a: INWARD boolnet;
    CONST
            year = 2001;
    TYPE
            color = (red, blue);
    VAR
            i: integer;
            c : color;
    PROCEDURE doit(k : integer);
            BEGIN
            writeln(k, year);
            END;
    BEGIN
            writeln('hello');
    END;
```

would turn into:

```
PROCEDURE foo(external : z9_comp_base);
   BEGIN
   LABEL 9002, 9999;
   CONST year = 2001;
   TYPE
      color = (red, blue);
      z9_comptype_state = RECORD
         a : z9_net_base_ptr;
         i : integer; c : color;
         END
   VAR (*type coercion*)
      internal : RECORD
         CASE boolean OF
            true: (actual : ^ z9_comptype_state);
            false : (fake : integer_ptr);
            END;
   PROCEDURE doit(k : integer);
      BEGIN
      WITH internal.actual^ DO
         BEGIN
         writeln(k, year);
         END;
      END;
   BEGIN
   internal.fake := external^.internal.users; (* COERCION*)
   WITH internal.actual^ DO
      BEGIN
      CASE external^.delta OF
         0: BEGIN (*initialize storage areas *)
            new(external); (* sim. sys. control area *)
            new(internal); (* user defined component variables area*)
            external^.internal.users := internal.fake;
            external^.delta := 1;
            (*read actual parameters to component here*)
            GOTO 9999;
            END
         2: GOTO 9002;
         1: BEGIN (* user suplied main body*)
            z9_isolate(external);
            (*code to hook up subprocesses goes here*)
            writeln('hello');
            external^.delta := 2;
            GOTO 9999;
            9002: writeln('good bye');
            external^.delta := 9999;
            GOTO 9999;
            END (* case 1*)
         END; (*CASE*)
      END; (*WITH*)
   9999:;
   END; (* PROCEDURE*)
```

6.5 Generating Code to Support Net Types

Net type definitions generate several separate pieces of Pascal code. First the net type names turn into fields in a Pascal variant record type called z9_net_value. The variant record is discriminated by the number of the net type. Secondly, two procedures are generated for manipulating net values. The first is called z9_create_net_value, and it is responsible for allocating storage for new nets. The second is z9_net_handler. This accepts pointers to two nets and determines if they have the same value. If the values are identical it sets a flag to false and returns. If not, it sets the flag to true and assigns one to the other. Thus the ADLIB code:

```
NETTYPE
      boolnet = boolean;
      real_net = real;
```

would generate one record and two procedures. The record would look like:

```
z9_net_value = RECORD
      fieldnum: integer;
      CASE nettype_num : integer OF
            1: (boolnet : boolean );
            2: (real_net : real);
            END; (* CASE *)
      END; (*RECORD*)

      z9_net_value_ptr = ^z9_net_value;
```

One procedure is used to allocate new members:

```
PROCEDURE z9_create_net_value(nettype_num: integer ;
 VAR result : z9_net_value_ptr);
   BEGIN
   CASE nettype_num OF
      BEGIN
      1: new(result,1);
      2: new(result,2);
      END (*CASE*)
   END; (* PROCEDURE *)
```

and the other procedure is used to manipulate them:

```
PROCEDURE z9_net_handler(net1, net2:z9_net_value_ptr;
 VAR change: boolean);
   BEGIN
   CASE net1.nettype_num OF
      1: BEGIN
         IF (net1^.boolnet = net2^.boolnet) THEN
            change := true
         ELSE
            BEGIN
            change := false;
            net1^.boolnet := net2^.boolnet;
            END
         END;
      2: BEGIN
         IF (net1^.real_net = net2^.real_net) THEN
            change := true
         ELSE
            BEGIN
            change := false;
            net1^.real_net := net2^.real_net;
            END
         END;
      END (*CASE*)
   END;
```

In the general case the code may be slightly more complicated if one or more of the net types is a compound type (*i.e.* an array or record). This is because it is legal for a component to update one element in a structured net; the semantics of ADLIB state that the others must not be altered*. Thus the standard Pascal array or record comparisons and assignment operations cannot always be used. To support ADLIB's semantics, each update to a compound net carries with it an extra integer field. This extra field indicates which user-declared field, or array element, in the record is the one to be compared or updated. Support for this includes a second level of CASE statements in the net value handling subroutines above.

6.6 Samples of ADLIB Preprocessor Transformations

Here is list of some of the other transformations needed for various constructs.

1. Pascal global constants, types, and variable declarations: No transformations are required for these, except to check for ADLIB reserved words and verify correct type usage.

2. Pascal expressions not involving nets: These are passed through unchanged, but checked for type.

3. Declarations of ports inside component types: These are numbered internally and appear as fields in the user_data record associated with the component type procedure.

4. Expressions referring to ports inside component types: First the direction of the port is checked to verify that reading the port is allowed (*i.e.* INWARD, INTERNAL or BOTHWAYS). The expression is transformed into a new expression, consisting of the name of the port, dereferenced, and with the field name instant appended (instant meaning "the instantaneous value"), and the net type name attached. Thus if reset is an INWARD port of net type boolnet, the code:

        writeln(reset)

    is replaced by:

        writeln(reset^.instant.boolnet).

    The type of field instant must be appropriate to the context, as determined by the net type declared for the port. Note that this allows the built-in Pascal procedure writeln to generate the

---

* Theoretically, net types of any level of nested structured data types should be supported. However, no implementation has ever supported more than one level.

correct code to print out a boolean value.

5. Net update statements (ASSIGN and part of TRANSMIT): The preprocessor first verifies that the direction of the target port is correct (*i.e.* OUTWARD, INTERNAL, or BOTHWAYS). Then it generates a Pascal block (delimited by BEGIN END) containing two statements. The first is a Pascal assignment statement that takes the user-supplied expression and loads it into an update buffer supplied by the simulation environment. If the target port has a compound net type, and only one subfield within the net is to be updated, an integer constant is specified in the update block, as discussed above. Next a Pascal statement is generated that calls the simulation support subroutine assign. The assign subroutine takes the buffer and links it into the event queue, then provides a fresh buffer for the next ASSIGN statement. The assign subroutine takes as arguments a pointer to the net to be updated and the time of the update, as determined by the timing control clause (DELAY or SYNC).

6. DETACH statements: These turn into an assignment that saves the point of execution (as an integer), a GOTO statement that branches out of the component type procedure, and a label that can be used for restoring the state. *E.g.*

```
a := b + 1;
DETACH;
write('hello');
```

turns into:

```
a := b + 1;
BEGIN
external^.delta :=  27;
GOTO 9999;
9027:
END;
write('hello');
```

7. Subroutine definitions outside of component type definitions: If a port name is passed by value as a formal argument to a subroutine, its type must be replaced in the argument list by the type of the underlying net. The code inside the subroutine that accesses the parameter then needs no further manipulation. If the port is passed

by reference (VAR), then the type must be replaced by the universal net reference type `z9_net_base_ptr` in the argument list. Wherever such arguments appear in expression they must be expanded to the name of the net and dereferenced with the net type name appended, as discussed above. Only if the net is passed by reference can it be updated inside a subroutine.

8. Subroutines declared inside component types: In addition to the above transformations, these must have a Pascal WITH statement enclosing them to provide access to the variables local to the component type definition.

9. FOR loops inside component types: Because some Pascal compilers allow only local scalar variables to be the subject of iteration in FOR loops, and there are no local scalar variables in the component type procedures, ADLIB FOR loops must be rewritten as WHILE loops in Pascal. Thus

```
VAR
i : integer;

FOR i:= 1 TO 5 DO
        BEGIN
        DETACH;
        j := j * 4;
        END
```

becomes:

---

* Although it is efficient, this technique can result in scoping bugs.

```
BEGIN
    WITH external^.internal.users
        BEGIN
            BEGIN (*for loop code*)
            i := 1;
            WHILE i <=5 DO BEGIN
                BEGIN
                    external^.delta := 11;
                    GOTO 9999;
                    9011:
                END; (*detach code*)
                j := j * 4;
                i := i + 1;
                END; (*WHILE*)
            END; (*for loop code*)
        END; (*with*)
    END;
```

10. **Sensitize** and **desensitize**: These predeclared ADLIB subroutines translate into calls on a support subroutine, with an argument that points to the *receptor* that controls the interface between the net and the component. This is discussed in the section on simulator data structures below.

11. **WAITFOR** <expression> **CHECK** <list> statements: These translate into a sequence of operations. The first is a call on the support routine **z9_isolate**, which desensitizes the component to all its input nets. Next, a series of calls on **sensitize** is generated, one for each net listed in <list>. Third, code is generated that tests <expression>. If the expression evaluates to **false**, the program branches out again. For example, the ADLIB code:

```
c := b + 1;
WAITFOR NOT a CHECK a;
write('hello');
```

turns into:

```
c := b + 1;
    BEGIN (*waitfor*)
    z9_isolate(external);
    sensitize(a);
    external^.delta :=  28;
    GOTO 9999;
    9028:
    IF NOT (NOT a) THEN GOTO 9999;
    END; (*waitfor*)
write('hello');
```

At this point it should be obvious why DETACH is more efficient than WAITFOR: the overhead of isolating and resensitizing a set of nets is completely avoided.

12. WAITFOR statements with DELAY: These are almost identical to those with CHECK except that the call on sensitize is replaced by the appropriate call on the scheduling package.

13. Subprocesses (TRANSMIT and UPON): There have been several approaches to supporting these; one turns them into subroutines, the other changes them into groups of executable statements within the body of the component type procedure. The second approach seems to work better. Thus the UPON subprocesses

```
UPON a CHECK a DO
     BEGIN
     b := c;
     END;
UPON reset = a  CHECK reset, a DO
     BEGIN
     b := d;
     END;
```

turn into:

```
9901:
      IF NOT (a) THEN GOTO 9999;
      b := c;
      GOTO 9999;
9902:
      IF NOT (reset = a) THEN GOTO 9999;
      b := d;
      GOTO 9999;
```

The sensitizing or scheduling of subprocesses is done at simulation time zero, and also by the user through calls on the inhibit and permit routines. These manipulate receptor records which control the relation between the component and the nets driving it. The mechanism is the same as the one that controls the main body of the component type, except that the data structures are marked with a field designating which subprocess is to be activated. TRANSMIT is almost identical, except that net update code is generated automatically.

14. CASE, WHILE and REPEAT loops: These pass through unchanged.

15. CLOCK definitions: These are purely preprocessing time entities. The preprocessor records the period and number of phases of each definition, and prints this information as constants in the code that implements the SYNC expressions. This code makes use of the Pascal MOD operator to find the next time, greater than the current time, that the clock will be at the specified phase.

16. Component type parameters: The preprocessor moves component type parameters into the ordinary variables within the component type procedure. It also generates a call on the Pascal support subroutine readln to initialize them. This call is executed at simulation time zero by the support code. The simulator is set up so that the structure control file is positioned correctly and the right information for each component instance will be on the next line to be read*. This is done for all components with parameters.

---

* On the DEC-10 version this makes use of the Pascal compiler's ability to read quoted strings, and all the scalar types, including enumerated types, quoted characters, and so forth.

6.7 Discussion: Global Variables, Scoping and Information Hiding

The ADLIB preprocessor takes ADLIB global variables and maps them directly into global variables in the Pascal code produced. These can be used during simulation to communicate freely from component to another, and between the main body of the ADLIB file and the components. As was discussed in Chapter 3, global variables are potentially dangerous, since they allow unstructured communication from one component to another. And some programming tasks cannot be done without them.

6.8 Simulation

The simulator itself is built out of the Pascal code described above, plus a fixed run-time support package that need not be recompiled for each ADLIB file.

6.9 Basic Data Structures

The run-time event processing is designed to minimize the overhead per event and eliminate busy waiting. The objective is that the efficiency should remain relatively constant over a wide range of simulation sizes. To implement this, the support environment contains data structures with a large number of pointers, some of which, unfortunately, are redundant, (in the sense that they could be derived from other data). While redundancy is normally something to be avoided, it is necessary here so that the simulation code can reach directly from one object to the objects connected to it; no searches and only a few array accesses are ever required. This will be illustrated for the three basic types of operations that occur during simulation: accessing and assigning to nets, stimulation of components, and event scheduling.

6.10 Accessing and Updating Nets

Each net used in the simulation requires a set of records to be allocated and linked together. The main record is called net_base, and every net has exactly one of these. The net_base is pointed to by a z9_net_base_ptr inside the components that interact with it. The net base record, in turn, points to a z9_net_value record called instant which contains its value at any instant of simulated time. The net_base also has a pointer to a z9_net_value record called update, which is a buffer used by ADLIB net-update statements. As discussed above, the instant field is used by components to read the values of nets, and the update buffer is used to update them. As components run, they execute net update statements (ASSIGN and TRANSMIT). Each of these loads new data into the update buffer, and then the assign() procedure stores it away on the event list. For each net type defined by the user, a pool of spare **z9_net_value** records is maintained, waiting for reuse. This structure is illustrated below:

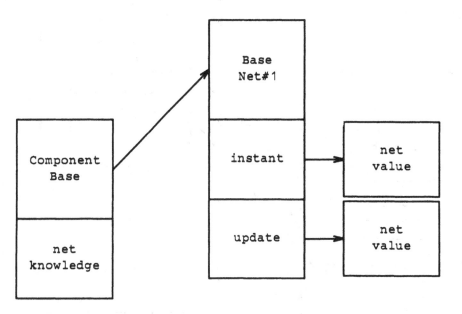

Figure 6-2. How a Component Updates a Net

6.11 Nets Stimulating Components

Each of the **z9_net_base** records has a list of receptor records. These mediate the stimulus-driven nature of the simulation. Each is pointed to directly by the component and is linked into a list of receptors for each net. Thus the component can get to the receptor quickly for **sensitize** and **desensitize** operations, and the net can get quickly to all components that are sensitive to it. All the receptors for a given component are themselves threaded together, so that the component can isolate itself from all its nets quickly with the **z9_isolate** subroutine. This is shown in Figure 6-3.

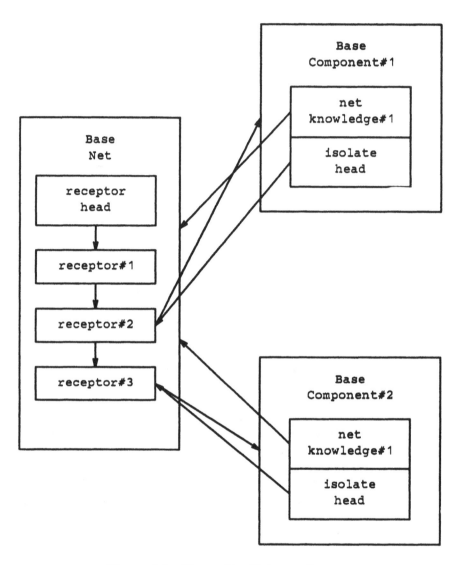

Figure 6-3. How a Net Drives a Component

6.12 Event Scheduling

Since ADLIB is an event-based language, it is not surprising that simulation control centers around an event list. This consists of a singly-linked list of events, each with a different and ascending time stamp. Single linking is acceptable for events because they are never deleted

except from the head of the list. Connected to each event is a list of
net-update and component-awakening records. Because the semantics of
the language requires that net updates and component awakenings be
deletable, this second list is doubly linked, thus making it easy to remove
individual updates. This is shown below:

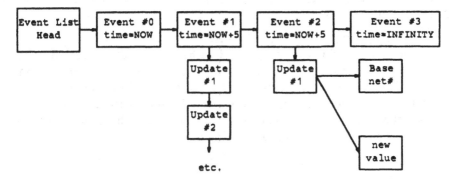

Figure 6-4. Simulation Event List

6.13 Before Time = Zero

All these data structures are dynamically allocated when simulation is
started. The process works as follows: First, the design's structure is read
in from the Structure Description Language (SDL) file. For each net
mentioned in the SDL, a record of the appropriate type is allocated on
the heap using the routine z9_create_net_value, (which was created
by the preprocessor.) Next, for each component included in the SDL a
new component base (comp_base) record is created. The component
type procedure is called to allocate the data area needed by the
component. The component type procedure also reads any actual
parameters to the component at this time. Finally, the components and
nets are threaded together. This is accomplished by taking the SDL
structure information and finding pairs of nets and components. For
example, net number 1 may be connected as the third port inside
component nand1. The routine z9_hookup is called that reaches inside
the components state record and performs the appropriate linkages
(strong typing must again be broken). This is the point where the
receptor records are allocated and linked. After all the linking and new
operations are completed, all the components are invoked once more to
allow any user supplied initialization code to be run. When this is done,
control passes to the simulation monitor for user interaction.

6.14 The Basic Event Cycle and its Efficiency

The basic cycle of simulation outlined above is more complex than that of many simulators because of its high degree of user programmability and extensibility. The data structures were designed to compensate for this complexity by letting the basic cycle be as time efficient as possible.

The data structures chosen provide quick access for the most basic ADLIB operations. Specifically:

1. Invoking a component body: Because Pascal does not allow pointers to subroutines, this is implemented with a CASE statement that calls the desired component type body. The efficiency of this depends on how well the underlying Pascal compiler implements CASE statements. Once called, there is a small fixed overhead associated with the WITH statement (in some compilers) and the CASE ... GOTO to implement the multi-programming environment.

2. Invoking a subprocess: This is just a variation of invoking a component body.

3. Access to local variables stored within a component instance: These are accessed as fields in the state record for the component using a WITH operation. Some Pascal compilers optimize the WITH operation using registers, making this as efficient as accessing ordinary local variables.

4. Reading or assigning to a net: Reading a net is accomplished in constant time just by dereferencing a pointer. Net update statements also take constant time, except possibly for the time required to insert an update into the event list.

5. Adding an event: As mentioned above, the original support environment used an ordinary linked list for the event list. Since the basic operation here is inserting into a sorted list, it can take time proportional to the number of distinct elements. However, since many digital systems are largely synchronous, the number of distinct times tends to be small, thus increasing efficiency. Also, a large percentage of the updates are made with delay 0, and thus are caught on the first event. In any case, there are many techniques such as time wheels, binary trees, *etc.* that can be used if this proves to be a bottleneck.

6. Finding all updates at a given time: These are available directly from the event record.

7. Finding all components sensitive to changes on a net: This is available directly from the receptors that hang off each net.

The use of these data structures results in a time of about 500 microseconds per event/update cycle on a DEC-10. With the exception of the event queue itself, there is little extra overhead for large simulations, which means that this number should be relatively stable.

In comparison with other simulation techniques, the structures defined here tend to achieve time efficiency at the cost of using more memory. However, the trend in machine design has been increasing the memory/MIP ratio over the last few years: today multiple megabyte memories are available on 1 MIP workstations. In view of this trend, this tradeoff seems appropriate.

6.15 Interactive Support

The run-time package must also include code that interacts with the user. Since ADLIB is capable of generating its own stimulus and evaluating its own state, it is possible to run with little interactive support. But at a minimum, it should allow the user to set the simulation time limit either by time or by number of events. A more sophisticated environment could allow the user to stop the simulation, examine and set the values of nets, and restart. A still more sophisticated environment, such as the one on DEC-10, makes use of the Pascal debugger to provide access to internal variables and to set breakpoints inside individual components.*

6.16 Alternative Structures

The original ADLIB support environment provided only *transparent* delay on nets. In a transparent delay through an inverter, for example, all the short duration transitions and glitches in the input are mirrored in the output. This happens in simulation even though a practical inverter has a finite bandwidth that would, in practice, stop any pulses shorter than a critical width. One alternative to the naive transport delay is the inertial delay model, wherein updates passing though an inverter shorter than a given duration are suppressed. This can be accomplished in simulation by making sure that extraneous net update records are deleted as new update records are entered into the event list. This would be difficult within the structures described above: a complete search of the event list might be necessary. A more efficient solution would be to link all the updates associated with each net, so that any new update could be

---

* The last is tricky, however, because which the user normally thinks of the code as existing within a single instance of a component, that same code is actually reused for all instances of the same component type. This causes the debugger to stop repeatedly as other instances are activated.

compared and redundant ones eliminated.

Other extensions to the language may require comparable enhancements to the simulation environment. One such enhancement, found in HHDL, is the ability for a component to identify which net awakened it directly. This requires that the index of the awakening net be stored in each component base. Another important feature available in HHDL is the ability to examine previous values of nets. This can be accomplished by storing one or more previous values on additional net_value fields on the net_base, along with the times of their updates. This data also makes it simple for a component to determine whether a particular net has been stable, rising, or falling.

6.17 Summary

The system described here represents a successful, but preliminary, approach to supporting the ADLIB language. It would be better to have a true compiler that could accept ADLIB source and produce optimal machine code, rather than depend on a Pascal compiler that must be both efficient and forgiving (in terms of type checking and GOTO's). But the preprocessor approach is more portable, has been adequate to demonstrate that the ADLIB language can be implemented, and that it can have performance competitive with less flexible hardware description languages.

CHAPTER 7

CONCLUSION

The first six chapters have described the ADLIB language and its present applications in some detail. This chapter will summarize the key features of ADLIB, assess how it fits into the world of CAD today, and discuss future prospects for hardware description languages in general, and ADLIB in particular.

7.1 Summary: The Essence of ADLIB

Although there are some differences in dialects of ADLIB, they all share the most important points:

- The underlying model is one of autonomous components communicating through their ports onto nets. The topology of the components and nets may be specified independently of the behavior of the components. This separation of structure and behavior allows the language to represent hierarchy effectively and efficiently. Furthermore, it allows the designer to modify the structural representation without recompiling the behavior description, and *vice-versa*.

- The datatype of the nets used in an ADLIB description are user-definable, and may range from low-level electrical representations (such as high_impedance), to individual bits, to complex records that model abstract concepts. This data abstractions facility allows ADLIB to model a wide range of hardware designs at many levels of abstraction. For example, a computer designer can model a control bus as a set of mnemonic values that match the instructions of the machine without having to describe the underlying bits. This net-type defining facility is perhaps the most important facility in ADLIB, since it means the user can tailor the language to fit the problem.

User-defined net types also have relevance to the definition of *multi-*

level simulation. Multi-level simulation is here defined as occurring if and only if a piece of information appears on two or more nets in different formats (net types), where one format is an abstract approximation of the other. Multi-level simulation benefits the designer, who can focus in on one part of the design, as well as the CAD system, because it can concentrate its resources on the most critical part of the problem.

In addition to their use in hardware description *per se*, ADLIB's user-extensible net types turned out to be useful for CAD developers doing other types of analysis, including fault simulation, timing verification, and symbolic simulation.

- Because ADLIB allows the user to define new types of nets, more information about a design's intent can be captured in machine-readable form. This makes it possible to verify this intent in a new way: ports on two components can be checked for compatibility when the structure specification indicates that they are connected by a net. This hardware modeling principle is directly analogous to the notion of *strong typing* found in programming languages.

- Another software concept that found its way into hardware modeling is the notion of *information hiding*. Here the parallel is between software modules and ADLIB hardware components, because in ADLIB, each component's function is defined only by the behavior that can be observed on its ports, not by any internal coding.

- The facilities in ADLIB are a superset of those in the general-purpose programming language Pascal, which gives the versatility needed to model complex hardware and software designs. In particular, ADLIB has been successfully used to model VLSI systems such as a Motorola 68000 microprocessor at several levels. This model, which includes timing, was developed in six staff-months, which would have been nearly impossible without the kind of extensive programming facilities that ADLIB provides via Pascal.

- ADLIB is a *process-oriented* simulation language, with *inter-process communication* facilities tuned to the problem of describing hardware. The basic model of a *process* in ADLIB is that of an autonomous component, with its own dedicated local variables and state. This part of the model is close to the notion of processes found in computer operating systems. But the ADLIB model for process communication is different than the one found in most operating systems: it is based on the use of nets for communication and control. Communication occurs when one component puts new information onto a net. Control is transferred when the receiving component is awakened by a change on a net, which is why ADLIB is described as a *stimulus-driven* environment. Because the ADLIB model is close to that of processes on a computer, it is easy to model high-level software facilities such as protocols. But because its inter-process communication facilities are analogous to hardware, low-level modeling is expedited too.

- Just as the stimulus-driven nature of ADLIB helps it to model asynchronous systems, the CLOCK and SYNC primitives expedite the modeling of synchronous systems, which is important because most intermediate-logical or register-transfer-level work is synchronous.

- Likewise, the ability for a component to be augmented with an unlimited number of subprocesses helps to capture concurrency within complex components, without the effort and overhead of altering the structure of the model.

Most importantly, ADLIB is capable of unambiguously specifying behavior over a wide range of abstraction levels, making it possible to use a single tool from the conception of a design through its completion.

7.2 Lessons Learned from the Original ADLIB

Over the years, several lessons have been learned from working with ADLIB. In particular, three of these stand out: the need for separate program development, for syntactic support for register-transfer operations, and for fully general support for net type definitions.

1. The original ADLIB language required that an entire set of component behaviors be compiled at once. This worked fine for one-person projects, but in industry, the vast majority of VLSI chips are designed by teams of engineers working in close consort. It is therefore critical that an HDL support separate compilation of design information, allow sharing of design data between engineers, and at the same time, allow some portions of a design to be protected or private.

2. Secondly, for certain stages of electronic design, (especially in the design of gate-array or standard-cell chips,) the register level of abstraction is the most important. A good hardware description language therefore needs an expressive, but terse facility to describe the basic operations at this level. In the original ADLIB, these facilities only were available only via Pascal subroutine calls. While this approach offered the same power found in most register-level description languages, it had significant deficiencies in readability and conciseness.

3. Finally, some implementations put restrictions on the Pascal types that could be used as ADLIB net types, or on how well they were supported at run time. These limitations proved highly annoying, because as users learned about net type definitions, they quickly took advantage of them in unexpected ways.

It was precisely these areas that received the greatest enhancements in HHDL/HELIX.

7.3 Trends in Hardware Description Languages

The CAD community is currently experiencing a number of dramatic changes, both technical and economic, and some of these are influencing the design of hardware description languages. On the one hand, as VLSI chips become larger, cheaper, and more automated, there is a need for higher-levels of description and design support. At the same time, a clear need is emerging for industry standards that would facilitate communication among design groups, fabrication facilities, and CAD tools, and would insure that investments in designs will not be lost as tools evolve.

This pressure for design standards has produced some contributions to the hardware description language area. In particular, the Engineering Design Interchange Format (EDIF) language, developed by a consortium of electronic and CAD companies, offers a potential solution for some hardware design transportation problems. It is particularly well suited to the problems of describing hierarchical structure and mask layout. In addition, the VHSIC Hardware Description Language (VHDL), developed under sponsorship by the United States Department of Defense (DoD), may become an industry standard for logic description and register-transfer level modeling, especially for projects associated with the DoD's Very High Speed Integrated Circuit (VHSIC) project.

7.4 The VHDL Language

The VHDL language is receiving substantial support from both the DoD and industry, and may turn out to be one of the most important hardware design languages to emerge in recent time. VHDL incorporates many of the features found in ADLIB, but at the same time includes several features which are unique, and worth discussing.

- Like some classical HDL's, VHDL combines the behavior and structure specifications into a single file. An interesting addition in VHDL though, is its ability to describe a combination of structural and RTL behavioral decomposition within a single design syntax. This intermediate stage allows a designer to combine logic expressions into the structure description. and may be a good match to the human process of functional design and specification in cases where the designer has not fully determined either the behavior or the structure, but wishes to mix-and-match the two.

- VHDL is based on Ada, which was also developed by the DoD. Ada is a large and powerful language developed for programming embedded military systems, and VHDL therefore naturally inherits much of that power. Of course, this leads to some difficulties, *e.g.* novices might confuse VHDL's notion of concurrency with Ada's. But such a base language also brings with it many advantages in terms of general programming features, portability of software, and user acceptance.

- The VHDL language is the result of a large and lengthy development and review procedure. One positive consequence of this is that the VHDL language designers were able to put quite a bit of effort into insuring that the semantic model in VHDL was well defined, especially the model for simulated time. As a result, the VHDL language, like Ada, should be tool independent.

If there is a weak spot in VHDL, it is probably in its facilities for modeling very high-level software entities. This is because, at the present time, VHDL Version 7.2 does not encorporate the process-model of simulation. Rather, each component in VHDL acts like a subroutine in that its state is lost after each execution. But this limitation is not inherent to VHDL, and since efforts are underway by the Institute of Electrical and Electronic Engineers (IEEE), and others, to enhance the VHDL language, any current deficiencies may be remedied in future releases.

7.5 Future Work

What does the future hold for ADLIB? It has already demonstrated its usefulness in several CAD areas beyond simulation, including fault analysis, behavior verification, and timing verification. But there is one area to which ADLIB has not been applied that may yet bear fruit: direct hardware synthesis. This is not to say that ADLIB/SDL has not been applied to building chips, since many chips have been built using the structure specified in an SDL file that had been paired with ADLIB for simulation. But this is only done after the design has been broken down into detailed structures. In order to take fuller advantage of the power of VLSI, it would be desirable to work with higher-level ADLIB code and generate custom hardware blocks to implement them directly. In recent years there has been significant interest in the automatic synthesis of specific hardware technologies, such as PLA's and micro-sequencers. It may be possible to extend this work by identify those ADLIB programming constructs that most conveniently map onto specific hardware entities, thus facilitating the translation of specifications directly into wire lists, masks, ROM patterns, and so forth. For example, some forms of TRANSMIT subprocesses should map directly into parallel logic engines. Whether such a technique would could be made practical remains an open issue. But in such an endeavor, the ability to do comprehensive simulation and verification on exactly the design that is to be synthesized could be used to simplify and accelerate the design process.

ADLIB SYNTAX

A.1 Low-Level Syntax

The basic format of ADLIB programs is patterned closely after Pascal. However, a few differences exist.

1. Identifiers may include the underscore "_", and the use of upper or lower case characters is insignificant.

2. In order to shorten the code, reduce programmer effort, and eliminate transcription errors, a user may "include" files into his or her ADLIB source. The syntax is (starting in column 1):

    ```
    %include "<filename>"
    ```

    "Filename" must be a valid, unambiguous file name. Its syntax may depend on the operating system employed.

3. Comments are delimited by "(*" and "*)" and may be nested to any depth. (The original version of ADLIB also supported a convention where comments ran from "!" to the end of line.)

A.2 Summary of Operators

operator	operation	operand	result
Universal:			
ASSIGN	net assignment	expression,net,timing clause	
:=	assignment	any type except file	
Arithmatic:			
+(unary)	identity integer	same	
-(unary)	sign inversion		
+	addition		
-	subtraction		
*	multiplication		
DIV	integer division	integer	integer
MOD	modulus	integer	
/	real division	integer or real	real
Relational:			
=	equality	scalar,string,	boolean
<>	inequality	set or pointer	
<	less than	scalar or string	
>	greater than		
<=	less or equal	scalar or string	
	-or-		
	set inclusion	set	
>=	greater or	scalar or string	
	equal -or-		
	set inclusion	set	
Logical:			
NOT	negation	boolean	boolean
OR	disjunction		
AND	conjunction		
Sets:			
IN	set membership	scalar, and set	
+	union	any set type T	T
-	set difference		
*	intersection		

A.3 Predeclared Identifiers

The following are the standard, predefined identifiers in ADLIB. A user is free to use or redefine any of them, and implementors are at liberty to include additional predefined constants, types, variables, and subroutines wherever they might be useful. An asterisk (*) indicates those that are in ADLIB, but not Pascal.

1. Constants:

    `false, true, maxint`

2. Types:

    `bit*, boolean, char, integer, real,`
    `register*, text`

3. Nettypes:

    `semaphore`

4. Files:

    `input, output`

5. Functions:

    `abs, arctan, chr, cos, eof, eoln, exp, ln,`
    `odd, ord, pred, round, sin, sqr, sqrt,`
    `succ, trunc`

6. Procedures:

    `desensitize*, detach*, get, inhibit*,`
    `new, pack, page, permit*,`
    `put, read, readln, reset, rewrite,`
    `sensitize*, stopsim*, unpack, write,`
    `writeln`

A.4 Reserved Words

An asterisk indicates those that are in ADLIB, but not Pascal. A plus sign indicates DEC-10 Pascal extension.

AND
ARRAY
ASSIGN*
BEGIN
CASE
CHECK*
CLOCK*
COMPTYPE*
CONST
DEFAULT*
DELAY*
DIV
DO
DOWNTO
ELSE
END
EXTERN
EXTERNAL*
FOR
FILE
FORTRAN+
FORWARD
FUNCTION
GOTO
IF
IN
INWARD*
INTERNAL*
LABEL
MOD
NETTYPE*
nil
NOT
OF
OTHERS+
OUTWARD*
OR
PACKED
PHASE*

PROCEDURE
PROGRAM
RECORD
REPEAT
SET
SUBPROCESS*
SYNC*
THEN
TO
TRANSLATOR*
TRANSMIT*
TYPE
UNTIL
UPON*
VAR
WAITFOR*
WHILE
WITH

A.5 Backus-Naur Form

To describe the syntax, a modified Backus-Naur Form is used. Non-terminals are denoted by words such as "statement." Terminal symbols are written in CAPITAL LETTERS if they are reserved words, or enclosed in 'single quote marks' if they are not. Since there are two forms of quote character, one may be used to quote the other, i.e. "'" or '"'. Each syntactic rule (production) has the form:

$$S \rightarrow E;$$
$$\rightarrow E;$$

where S is a syntactic entity and E is a syntax expression denoting the set of sentential forms (sequences of symbols) for which S stands. If more than one expression is shown, then each is allowed. Comments are enclosed with "/*" and "*/".

The syntax for ADLIB is as follows:

```
program → MODULE 'ident' ';' modblock  ;
        → PACKAGE 'ident' ';' mblock '.'  ;

mblock → mblock_parts ;
       → mblock mblock_parts ;

mblock_parts → use_part ;
             → label_decl_part ;
             → const_def_part ;
             → type_def_part ;
             → nettype_decl_part ;
             → var_decl_part ;
             → proc_func_decl_part ;
             → compound_stat ;

modblock → modblock_parts ;
         → modblock modblock_parts ;

modblock_parts → use_part ;
               → time_unit_stat ;
               → const_def_part ;
               → type_def_part ;
               → nettype_decl_part ;
               → clock_def_part ;
               → comp_def_part ;
```

```
/*********** USE DECLARATIONS ***************/
use_part → USE use_list ';' ;

use_list → 'ident' ;
           → use_list ',' 'ident' ;

/*********** TIME UNIT STATEMENT ************/
time_unit_stat → TIMEUNITS 'integer' ';' ;

/*********** LABELS *********************/
label_decl_part → LABEL label_list ';' ;
label_list → 'integer' ;
→ label_list ',' 'integer' ;

/********** CONSTANTS ********************/
const_def_part → CONST const_def_list ';' ;
const_def_list → 'ident' '=' constant ;
→ const_def_list ';' 'ident' '=' constant ;

constant → unsigned_num ;
→ '+' unsigned_num ;
→ '-' unsigned_num ;
→ 'ident' ;
→ '+' 'ident' ;
→ '-' 'ident' ;
→ 'string' ;

unsigned_num → 'integer' ;
→ 'real' ;

/********* TYPES *********************/
type_def_part → TYPE type_def_list ';' ;

type_def_list → 'ident' '=' type ;
→ type_def_list ';' 'ident' '=' type ;

type → simple_type ;
→ structured_type ;
→ pointer_type ;

simple_type → scalar_type ;
→ subrange_type ;
→ 'ident' ;
```

scalar_type → '(' ident_list ')' ;

subrange_type → constant '..' constant ;

structured_type → array_type ;
 → record_type ;
 → set_type ;
 → reg_type ;

array_type → ARRAY '[' index_list ']' OF type ;

index_list → simple_type ;
 → index_list ',' simple_type ;

record_type → RECORD field_list END ;

field_list → fixed_part ;
 → fixed_part ';' variant_part ;
 → variant_part ;

fixed_part → record_section ;
 → fixed_part ';' record_section ;

record_section → ;
 → ident_list ':' type ;

variant_part → CASE tag_field OF variant_list ;

tag_field → 'ident' ;
 → 'ident' ':' tag_type ;

tag_type → 'ident' ;

variant_list → variant_section ;
 → variant_list ';' variant_section ;

variant_section → case_label_list ':' '(' field_list ')' ;
 → ;

set_type → SET OF simple_type ;

reg_type → REGISTER '[' subrange_type ']' ;
 → REGISTER '[' 'ident' ']' ;

pointer_type → '^' 'ident' ;

/********* NETTYPES *************************/
nettype_decl_part → NETTYPE nettype_decl_list ';' ;

nettype_decl_list → 'ident' '=' type ;
                  → nettype_decl_list ';' 'ident' '=' type ;

/********* VARIABLES *********************/
var_decl_part → VAR var_decl_list ';' ;

var_decl_list → var_list ;
              → var_decl_list ';' var_list ;

var_list → ident_list ':' type ;

/********* CLOCKS ***************************/
clock_def_part → CLOCK clock_def_list ';' ;

clock_def_list
 → 'ident' '(' 'integer' ',' 'integer' ')' clock_def ;
 → clock_def_list ';' 'ident' '('
  'integer' ',' 'integer' ')' clock_def ;

clock_def → ;
 → DEFAULT ;

/********* PROCEDURES/FUNCTIONS ************/
proc_func_decl_part
 → PROCEDURE 'ident' formal_parms ';' block ';' ;
 → FUNCTION 'ident' formal_parms ':' 'ident' ';' block ';' ;
 → FUNCTION 'ident' ';' block ';' ;

block → block_parts ;
 → block block_parts ;

block_parts → label_decl_part ;
 → const_def_part ;
 → type_def_part ;
 → var_decl_part ;
 → proc_func_decl_part ;
 → compound_stat ;
 → EXTERN ;
 → FORWARD ;

→ FORTRAN ;

formal_parms → ;
  → '(' formal_parm_sec_list ')' ;

formal_parm_sec_list → formal_parm_sec ;
  → formal_parm_sec_list ';' formal_parm_sec ;

formal_parm_sec → ident_list ':' 'ident' ;
  → VAR ident_list ':' 'ident' ;

/*********** COMPONENT TYPES **********************/
comp_def_part
  → COMPTYPE 'ident' formal_parms ';' cblock ';' ;
  → TRANSLATOR 'ident' formal_parms ';' cblock ';' ;

cblock → cblock_parts ;
  → cblock cblock_parts ;

cblock_parts → default_part ;
  → net_ref_part ;
  → label_decl_part ;
  → const_def_part ;
  → type_def_part ;
  → var_decl_part ;
  → proc_func_decl_part ;
  → subp_decl_part ;
  → compound_stat ;

default_part → DEFAULT default_list ';' ;

default_list → 'ident' '=' constant ;
         → default_list ';' 'ident' '=' constant ;

net_ref_part → INWARD net_ref_list ';' ;
  → OUTWARD net_ref_list ';' ;
  → BOTHWAYS net_ref_list ';' ;
  → INTERNAL net_ref_list ';' ;

net_ref_list → net_list ;
  → net_ref_list ';' net_list ;

net_list → ident_list ':' 'ident' ;

subp_decl_part → SUBPROCESS subp_decl_list ;

subp_decl_list → subp_decl ;
              → subp_decl_list subp_decl ;

subp_decl → 'ident' ':' TRANSMIT
expression qualifier TO variable

timing_clause ';' ;
→ 'ident' ':' TRANSMIT

expression qualifier TO variable ';' ;
              → 'ident' ':' UPON

expression qualifier DO statement ';' ;

qualifier → check_list ;
→ timing_clause ;

timing_clause → DELAY expression ;
→ ; /* null timing clause*/
→ SYNC ;
→ SYNC 'ident' ;
→ SYNC 'ident' PHASE expression ;
→ SYNC PHASE expression ;

check_list → CHECK check_net_list ;

check_net_list → net_id ;
→ check_net_list ',' net_id ;
/********** STATEMENTS *******************/
statement → unlabelled_stat ;
→ 'integer' ':' unlabelled_stat ;

unlabelled_stat → ;
→ structured_stat ;
→ assign_stat ;
→ proc_stat ;
→ goto_stat ;
→ netassign_stat ;
→ waitfor_stat ;
→ detach_stat ;

assign_stat → variable_list ':=' expression ;

```
variable_list → variable  ;
              → variable_list ',' variable  ;

variable → 'ident'  ;
→ variable '[' indexes_list ']'  ;
→ variable '.' 'ident'  ;
→ variable '^'  ;
→ variable '[' expression '..' expression ']';

indexes_list → expression ;
→ indexes_list ',' expression ;



expression → simple_exp ;
→ simple_exp rel_operator simple_exp ;

rel_operator → '=' ;
→ '<>' ;
→ '<' ;
→ '<=' ;
→ '>=' ;
→ '>' ;
→ IN  ;

simple_exp → term ;
→ '+' term ;
→ '-' term  ;
→ simple_exp add_operator term  ;

add_operator → '+' ;
→ '-' ;
→ OR  ;

term → factor ;
→ term mult_operator factor  ;

mult_operator → '*'  ;
→ '/' ;
→ DIV  ;
→ MOD  ;
→ AND  ;
→ '//' ;
```

factor → variable ;
 → '(' expression ')' ;
 → 'ident' '(' expression_list ')' ;
 → '[' ']' ;
 → '[' element_list ']' ;
 → NOT factor ;
 → 'string' ;
 → nil ;
 → unsigned_num ;

element_list → element ;
 → element_list ',' element ;

element → expression ;
 → expression '..' expression ;

proc_stat → 'ident' ;
 → 'ident' '(' expression_list ')' ;

goto_stat → GOTO 'integer' ;

structured_stat → compound_stat ;
 → cond_stat ;
 → repetitive_stat ;
 → with_stat ;

compound_stat → BEGIN statement_list END ;

statement_list → statement ;
 → statement_list ';' statement ;

cond_stat → IF expression THEN statement ;
 → IF expression THEN statement ELSE statement ;
 → CASE expression OF case_element_list otherwise_clause END ;

case_element_list → case_element ;
 → case_element_list ';' case_element ;

case_element → ;
 → case_label_list ':' statement ;
case_label_list → constant ;
 → case_label_list ',' constant ;
otherwise_clause → ;
 → OTHERWISE statement ;

repetitive_stat → WHILE expression DO statement ;
→ REPEAT statement_list UNTIL expression ;
→ FOR 'ident' ':=' expression TO expression DO statement ;
→ FOR 'ident' ':=' expression DOWNTO expression DO statement ;

with_stat → WITH rec_var_list DO statement ;

rec_var_list → variable ;
→ rec_var_list ',' variable ;

netassign_stat
→ ASSIGN indexes_list TO net_id_list timing_clause;
→ ASSIGN indexes_list TO net_id_list ;
→ ASSIGN TRANSPORT indexes_list TO net_id_list timing_clause ;
→ ASSIGN TRANSPORT indexes_list TO net_id_list ;

waitfor_stat → WAITFOR check_list ;
→ WAITFOR timing_clause ;
→ WAITFOR expression qualifier ;

detach_stat → DETACH ;

/*********** MISCELLANEOUS *****************/
ident_list → 'ident' ;
→ ident_list ',' 'ident' ;

net_id_list → net_id ;
→ net_id_list ',' net_id ;
net_id → 'ident' ;

expression_list → expression io_length ;
→ expression_list ',' expression io_length ;
io_length → ;
→ ':' 'integer' ;
→ ':' 'integer' ':' 'integer' ;

A.6 Comments on Syntax

In the syntax for "timing_clause," if no expression is present, the expression "0" is assumed. If no 'ident' is present after SYNC, the default clock is used.

The DEC-10 version has a sightly different syntax for "case_element." Instead of OTHERWISE, it uses "OTHERS:." Its syntax for "program" allows an asterisk to appear after file names, which indicate that the file should not be rewritten. It also has a slightly different syntax for parameters to function parameters. The original version of ADLIB also requires names for subprocesses and supports only check lists for controlling them.

PACKAGES

One objective indesigning ADLIB was that the language itself should remain small, but at the same time be easily extensible, both by the user or the CAD implementor. In the case of the implementor, this is accomplished by supplying additional predeclared data types and subroutines, which is painless to the user because it does not imply any syntax changes or extensions, and does not require the user to learn about features not relevant to his or her own area of work.

Users, on the other hand, can enhance the software environment in another way by creating packages. This appendix describes one of such package, called rndpak, which provides a set of random number generators with various distributions and other facilities. In order to access a package, the user must the add the headings of its subroutines into his or her source program. This can accomplished with the USE statement, as in:

USE regpack;

B.1 Rndpak

Rndpak is a set of random number generators useful for stochastic simulations. They are listed below.

1. PROCEDURE rndset(newseed : integer);

Rndset resets the internal random number generator mechanism using newseed. It is only necessary to call rndset if multiple simulation runs are to be performed using different random inputs.

2. `FUNCTION rnd01 : real;`

`Rnd01` produces a random number between 0.0 and 1.0 by using one random number generator to scramble the results of another, thereby lowering the autocorrelation.

3. `FUNCTION rndnexp(lambda : real) : real;`

`Rndnexp` returns a number drawn from the negative exponential distribution with mean and standard deviation 1.0/`lambda` (`lambda` must be positive).

4. `FUNCTION rnderlang(lambda : real; k : integer) : real;`

`Rnderlang` returns a number drawn from the Erlang distribution with mean (1/`lambda`) and standard deviation 1/(sqrt(k)*`lambda`). (Minimum `k`=1, higher `k` makes for a tighter distribution.)

5. `FUNCTION rndnormal(mean, variance : real) : real;`

`Rndnormal` returns a number drawn from the Normal distribution with the `mean` and `variance` specified. The distribution is approximated by summing 32 uniformly distributed random values.

6. `FUNCTION rndint(low, high : integer) : integer;`

`Rndint` produces an integer evenly distributed among the numbers from low to and including high.

7. `FUNCTION rnddraw(p : real) : boolean;`

`Rnddraw` returns true with probability p.

8. `FUNCTION rnduniform(low, high : real) : real;`

`Rnduniform` produces a random real number uniformly distributed between low and high. (Note that the probability of returning a value exactly equal to low or high is vanishingly small.)

Data Analysis Facility

9. `PROCEDURE rndhisto(data : real; command : integer);`

`Rndhisto` collects, analyzes and plots a random variable. It produces a histogram automatically scaled to the width of the paper. Maximum number of bins=200. Its commands are as follows:

0	reset all tallies, and data values
1	set high limit to data (default = 10.0)
2	set low limit to data (default = -10.0)
3	set number of bins to "data (default = 20)
4	accept data as a point to be plotted
5	plot results in file output
6	plot results at tty
7	set paper width to data
8	reset all parameters to default values

The print out includes: the number of points, the value of the highest and lowest points, the number of points out of range high and low (if any), the mean, variance, sum, standard deviation, sum of squares, and the auto covariance and autocorrelation of adjacent terms. The following trivial program shows an example of its use:

```
PROGRAM x;
VAR i : integer;
FUNCTION rnderlang(lambda:real;k:integer):real; EXTERN;
PROCEDURE rndhisto(data:real;command: integer);EXTERN;
   BEGIN
   FOR i := 1 TO 1000 DO
     rndhisto(rnderlang(3.0,4),4);
   rndhisto(0.0,5);
   END.
```

This produces the output:

```
-9.50000     0>
-8.50000     0>
-7.50000     0>
-6.50000     0>
-5.50000     0>
-4.50000     0>
-3.50000     0>
-2.50000     0>
-1.50000     0>
-5.00000E-01 7>
 5.00000E-01 134>XXXXXXXXXXXXXX
 1.50000     297>XXXXXXXXXXXXXXXXXXXXXXXXXXXXXXXXXX
 2.50000     262>XXXXXXXXXXXXXXXXXXXXXXXXXXXXXX
 3.50000     154>XXXXXXXXXXXXXXXXXX
 4.50000     82>XXXXXXXXX
 5.50000     32>XXX
 6.50000     17>X
 7.50000      8>X
 8.50000      6>
 9.50000      0>
NUM POINT=      1000 LOWVAL= 1.826466932E-01
HIGHVALUE= 1.178417652E+01
1 POINT(S) WERE TOO HIGH 0 POINT(S) WERE TOO LOW
MEAN= 2.976368099    VARIANCE= 2.365632474
SUM= 2.976368099E+03
SUMSQ= 1.122203394E+04 SUM PROD= 8.854801416E+03
STD DEV= 1.538061276
AUTO COVARIANCE=-6.792426109E-03
AUTOCORRELATION=-2.871293902E-03
```

TRANSLATOR EXAMPLES

As discussed in Chapter 4, multi-level simulation between components using different net types to represent the same information can be accomplished with the use of translators. This appendix contains two examples of such translators, the first one being the one used in Chapter 4 to convert between an alu function and a group of independent nets, and the second one second being two-value logic into multi-value logic converter.

```
TRANSLATOR functolog;
  INWARD f:alu_function;
  OUTWARD i5,i4,i3:logtype;

  PROCEDURE update(i5val,i4val,i3val:logtype);
    BEGIN
    ASSIGN i5val TO i5;
    ASSIGN i4val TO i4;
    ASSIGN i3val TO i3;
    END;

  SUBPROCESS
    UPON true CHECK f DO
      CASE f OF
        fadd:   update(lo,lo,lo);
        fsubr:  update(lo,lo,hi);
        fsubs:  update(lo,hi,lo);
        flor:   update(lo,hi,hi);
        fand:   update(hi,lo,lo);
        fnotrs: update(hi,lo,hi);
        fexor:  update(hi,hi,lo);
        fexnor: update(hi,hi,hi);
        END;
  BEGIN
  END;
```

```
NETTYPE multinet=(x, u, d, l, h, s);

(*converts two value logic to multi-value*)
(* when the input net changes, it first assigns
** the going up (u) or going down (d) value.
** When the input net stablizes, it assigns
** the high (h) or low (l) values*)
TRANSLATOR tomulti;
INWARD
      a : boolnet;
EXTERNAL
      out : multinet;
INTERNAL
      int : boolnet;
SUBPROCESS
      checker : UPON a = int CHECK int DO
            IF a THEN ASSIGN h TO out ELSE assign 1 TO out;
BEGIN
ASSIGN initval TO out;
WAITFOR DELAY 0.0;
permit(checker);
sensitize(a);
WHILE true DO BEGIN
   DETACH;
   IF a THEN ASSIGN u TO out DELAY 0.0
   ELSE ASSIGN d TO out DELAY 0.0;
   ASSIGN a TO int DELAY 1.0;
   END;
END;
```

SYMBOLIC SIMULATION
SUPPORT ROUTINES

The following routines are used to support the symbolic simulation capabilities, as described in Chapter 5.

```
(* dynamic allocation and garbage collection *)
PROCEDURE newe(VAR e:eptr);
  BEGIN
  IF fsh = nil THEN NEW(e)
  ELSE
    BEGIN
    e:=fsh;
    fsh:=e^.left
    END;
  WITH e^ DO
    BEGIN
    e^.operator:=boole;
    e^.left:=nil;
    e^.right:=nil
    END
  END;

PROCEDURE disposee(VAR e:eptr);
  BEGIN
  e^.left:=fsh;
  fsh:=e;
  e:=nil;
  END;
```

```
PROCEDURE vari(n:alph):eptr;
  VAR e:eptr;
  BEGIN
  newe(e);
  WITH e^ DO
    BEGIN
    e^.operator:=xvar;
    e^.name:=n;
    END;
  vari:=e;
  END;

(* return constant boolean symbolic expression, true or false *)
FUNCTION bc(bval:BOOLEAN):eptr;
  BEGIN
  IF bval THEN bc:=symbtrue ELSE bc:=symbfalse
  END;

(* return constant integer symbolic expression with given value *)
FUNCTION ic(ival:INTEGER):eptr;
  VAR e:eptr;
  BEGIN
  IF (symplo <= ival) AND (ival <= symbhi) THEN e:=symbconst[ival]
  ELSE
    BEGIN
    newe(e);
    e^.operator:=integ;
    e^.i:=ival
    END;
  ic:=e
  END;

(* return true if symbolic expression is true *)
FUNCTION etrue(e:eptr):BOOLEAN;
  BEGIN
  IF e=nil THEN etrue:=false
  ELSE IF e^.operator=boole THEN etrue:=e^.b
  ELSE etrue:=false
  END;

(* return false if symbolic expression is false *)
FUNCTION efalse(e:eptr):BOOLEAN;
  BEGIN
  IF e=nil THEN efalse:=false
```

```
  ELSE IF e^.operator=boole THEN efalse:=NOT e^.b
  ELSE efalse:=false
  END;

(* return true if symbolic expression has value of 1 *)
FUNCTION eone(e:eptr):BOOLEAN;
  BEGIN
  IF e=nil THEN eone:=false
  ELSE IF e^.operator=integ THEN eone:=e^.I=1
  ELSE eone:=false
  END;

(* return true if symbolic expression has value of 0 *)
FUNCTION ezero(e:eptr):BOOLEAN:
  BEGIN
  IF e=nil THEN ezero:=false
  ELSE IF e^.operator=integ THEN ezero:=e^.i=0
  ELSE ezero:=false
  END;

(* return true if symbolic expression is a constant *)
FUNCTION econst(e:eptr):BOOLEAN;
  BEGIN
  IF e=nil THEN econst:=false
  ELSE econst:=(e^.operator=boole) OR (e^.operator=integ)
  END;

(* return true if two symbolic expressions are identical *)
FUNCTION esame(e1,e2:eptr):BOOLEAN;
  BEGIN
  IF (e1=nil) OR (e2=nil) THEN esame:=e1=e2
  ELSE IF e1^.operator=e2^.operator THEN
    CASE e1^.operator OF
      boole:esame:=e1^.b=e2^.b;
      integ:esame:=e1^.i=e2^.i;
      xvar: esame:=e1^.name=e2^.name;
      xodd,xord,xnot,xand,xor,times,xdev,xmod,plus,negate,
        minus,equal,ntequal,greater,less,grequal,lsequal:
          esame:=(e1^.left=e2^.left) and (e1^.right=e2^.right)
      END
  ELSE esame:=false
  END;

(* simplify an expression *)
```

```
PROCEDURE simplify(var e:eptr);
  VAR l,r,temp:eptr;
      ti:INTEGER;
      tb:BOOLEAN;
      op:ops;
  BEGIN
  IF e<>nil THEN
    BEGIN
    l:=e^.left;
    r:=e^.right;
    op:=e^.operator;
    IF op IN binaryops THEN
      BEGIN
      IF econst(l) AND econst(r) THEN
        BEGIN
        disposee(e);
        CASE op OF
          xand: tb:=l^.b AND r^.b;
          xor: tb:=l^.b OR r^.b;
          times: ti:=l^.i * r^.i;
          xdiv: ti:=l^.i DIV r^.i;
          xmod: ti:=l^.i MOD r^.i;
          plus: ti:=l^.i + r^.i;
          minus: ti:=l^.i - r^.i;
          equal: tb:=l^.i = r^.i;
          ntequal: tb:=l^.i <> r^.i;
          greater: tb:=l^.i > r^.i;
          less: tb:=l^.i < r^.i;
          grequal: tb:=l^.i >= r^.i;
          lsequal: tb:=l^.i <= r^.i
          END;
        IF op IN booleanops THEN
          BEGIN
          IF tb THEN e:=symbtrue ELSE e:=symbfalse
          END
        ELSE
          BEGIN
          IF (symblo <= ti) AND (ti <= symbhi) THEN e:=symbconst[ti]
          ELSE e:=ic(ti)
          END
      END
      ELSE (* reducible binaryop? *)
        BEGIN
        IF etrue(l) THEN CASE op OF
```

```
  xand,equal,lsequal: BEGIN disposee(e); e:=r END;
  xor,grequal:BEGIN disposee(e); e:=symbtrue END;
  less:BEGIN disposee(e); e:=symbfalse END;
  times,xdiv,xmod,plus,minus,ntequal,greater: END
ELSE IF etrue(r) THEN CASE op OF
  xand,equal,grequal: BEGIN disposee(e); e:=l END;
  xor,lsequal: BEGIN disposee(e); e:=symbtrue END;
  greater:BEGIN disposee(e); e:=symbfalse END;
  times,xdiv,xmod,plus,minus,ntequal,less: END
ELSE IF efalse(l) THEN CASE op OF
  xand,greater: BEGIN disposee(e); e:=symbfalse END;
  lsequal:BEGIN disposee(e); e:=symbtrue END;
  xor,ntequal,less:BEGIN disposee(e); e:=r END;
  times,xdiv,xmod,plus,minus,equal,grequal: END
ELSE IF efalse(r) THEN CASE op OF
  xand,less:BEGIN disposee(e); e:=symbfalse END;
  grequal:BEGIN disposee(e); e:=l END;
  times,xdiv,xmod,plus,minus,equal,lsequal: END
ELSE IF ezero(l) THEN CASE op OF
  times:BEGIN disposee(e); e:=symbconst[0] END;
  plus:BEGIN dispolsee(e); e:=r END;
  minus:BEGIN e^.operator:=negate; e^.left:=nil END;
  xand,xor,xdiv,xmod,equal,ntequal,greater,less,
    grequal,lsequal: END
ELSE IF ezero(r) THEN CASE op OF
  times:BEGIN disposee(e); e:=symbconst[0] END;
  plus,minus:BEGIN disposee(e); e:=l END;
  xand,xor,xdiv,xmod,equal,ntequal,greater,
    less,grequal,lsequal: END
ELSE IF eone(l) THEN
  BEGIN
  IF op=times THEN BEGIN disposee(e); e:=r END
  END
ELSE IF eone(r) THEN CASE op OF
  times,xdiv:BEGIN disposee(e); e:=l; END;
  xmod:BEGIN disposee(e); e:=symbconst[0] END;
  xand,xor,plus,minus,equal,ntequal,greater,
    less,grequal,lsequal: END
ELSE IF esame(l,r) THEN
  BEGIN
  IF (op=xand) OR (op=xor) THEN BEGIN disposee(e); e:=l END
  ELSE IF op In [greater,less,ntequal] THEN
    BEGIN disposee(e); e:=symbfalse END
  ELSE IF op In [grequal,lsequal,equal] THEN
```

```
          BEGIN disposee(e); e:=symbtrue END
        ELSE IF op=minus THEN BEGIN disposee(e); e:=symbconst[0] END
        END
      ELSE IF l<>nil THEN
        IF r<>nil THEN
          BEGIN
          IF ((l^.operator=xnot)AND esame(l^.right,r)) OR
             ((r^.operator=xnot)AND esame(r^.right,l)) THEN
            CASE op OF
              xand:BEGIN disposee(e); e:=symbfalse END;
              xor:BEGIN disposee(e); e:=symbtrue END;
              greater:BEGIN disposee(e); e:=l END;
              times,xdiv,xmod,plus: END
          ELSE IF ((l^.operator=negate)AND esame(l^.right,r)) OR
                  ((r^.operator=negate)AND esame(r^.right,l)) THEN
            IF op=plus THEN BEGIN disposee(e); e:=symbconst[0] END
          END
      END
    END
  ELSE (* unary operations *)
    BEGIN
    IF econst(r) THEN
      BEGIN
      disposee(e);
      CASE op OF
        xodd:tb:=odd(r^.i);
        xnot:tb:=NOT r^.b;
        negate:ti:=-r^.i;
        xord:ti:=r^.i;
        END;
      IF op IN booleanops THEN
        IF tb THEN e:=symbtrue ELSE e:=symbfalse
      ELSE IF (symblo <= ti)AND(ti <=symbhi) THEN e:=symbconst[ti]
      ELSE e:=ic(ti)
      END
    ELSE
      BEGIN
      IF op=xnot THEN
        IF r<>nil THEN
          IF r^.operator=xnot THEN BEGIN disposee(e); e:=r^.right END;
      IF op=negate THEN
        IF r<>nil THEN
          IF r^.operator=negate THEN BEGIN disposee(e); e:=r^.right ENI
      END
```

```
      END;
    IF e<>nil THEN
      IF ((e^.operator IN[binaryops+unaryops])AND(e^.right=nil))OR
         ((e^.operator IN[binaryops]        )AND(e^.left=nil)) THEN
            diposee(e)
    END
  END;

(* two routines used to create new copies of symbolic expressions *)
FUNCTION bx(op:ops; l,r:eptr):eptr;
  VAR e:eptr;
  BEGIN
  newe(e);
  WITH e^ DO
    BEGIN
    e^.operator:=op;
    e^.left:=l;
    e^.right:=r;
    END;
  simplify(e);
  bx:=e
  END;


FUNCTION ux(op:ops; r:eptr):eptr;
  VAR e:eptr;
  BEGIN
  newe(e);
  WITH e^ DO
    BEGIN
    e^.operator:=op;
    e^.left:=nil;
    e^.right:=r;
    END;
  simplify(e);
  ux:=e
  END;

(* assign one symbolic expression to another *)
PROCEDURE assignsymb(enew:eptr;var eold:symbolicnet;pd:integer);
  VAR e:eptr;
  BEGIN
  e:=eold;
  IF not esame(enew,e) THEN ASSIGN enew TO eold DELAY pd
  END;
```

APPENDIX E
BIBLIOGRAPHY

[ABR77] Abramovici, M., M. A. Breuer, K. Kumar, "Concurrent Fault Simulation and Functional Level Modeling," *Proceedings of the 14th Design Automation Conference*, New Orleans, 1977, pp. 128-137.

[AGR80] Agrawal, V.D., A.K. Bose, P. Kozak, H.N. Nham, E. Pacas-Skewes, "The Mixed Mode Simulator," *Proceedings of the 17th Design Automation Conference*, Minneapolis, 1980, pp. 626-633.

[AGR82] Agrawal, V.D., "Synchronous Path Analysis in MOS Circuit Simulation," *Proceedings of the 19th Design Automation Conference*, Minneapolis, 1980, pp. 626-633.

[ANL79] Anlauff, P. Funk, and P. Meinen, "PHPL - A New Computer Hardware Description Language for Modular Description of Logic and Timing, "*Proceedings of the 4th International Symposium on Computer Hardware Description Languages*, Palo Alto, Ca., Oct., 1979, pp. 124-130.

[ANT78] Antoniadis, D. A., S. E. Hansen, and R. W. Dutton, *Suprem II - A Program for IC Process Modeling and Simulation*, Tech. Report Number 5019-2, SEL78-020, Stanford Electronics Lab., June 1978.

[BAR77] Barbacci, M. et al., "An Architectural Research Facility - ISP, Descriptions, Simulations and Data Collection," *Proceedings of AFIPS NCC*, 1977, pp. 161-173.

[BAR79] Barbacci, M. R., G. E. Barnes, R. G. Cattell, D. P. Siewiorek, *The ISPS Computer Description Language*, CMU-CS-79-137, Departments of Computer Science and Electrical Engineering, Carnegie-Mellon University, 16 August 1979.

[BAR79-2] Barbacci, M. R., W. B. Dietz, L. J. Szewerenko, "Specification, Evaluation, and Validation of Computer Architectures Using Instruction Set Procesor Descriptions," *Proceedings of the 4th International Symposium on Computer Hardware Description Languages*, Palo Alto Ca., Oct., 1979, pp. 14-20.

[BAY79] Bayegan, H. M., O. Baadsvik, O. Kirkaune, "An Interactive Graphic High Level Language for Hardware Design," *Proceedings of the 4th International Symposium on Computer Hardware Description Languages*, Palo Alto, Ca., Oct. 1979, pp. 184-190.

[BEL71] Bell, C. G., A. Newell, *Computer Structures: Readings and Examples*, McGraw Hill, New York, 1971, pp. 22-33.

[BER71] Berry, D., "Introduction to Oregano," *Sigplan Notices*, Feb. 1971, pp. 171.

[BIR73] Birtwistle, G. M., O-J Dahl, B. Myhrhaug, K. Nygaard, *SIMULA BEGIN*, Auerbach Publishers, Inc., Philadelphia, Pa. 1973.

[BOS77] Bose, A. K., S. A. Szygenda, "Detection of Static and Dynamic Hazards in Logic Nets," *Proceedings of the 14th Annual Design Automation Conference*, New Orleans, La. June 1977, pp. 220-224.

[BRE76] Breuer, M. A., A. D. Friedman, *Diagnosis Reliable Design of Digital Systems*, Computer Science Press, Inc., Woodland Hills, Ca, 1976.

[BRY81] Bryant, R.E., "MOSSIM: A Switch-Level Simulator for MOS LSI", *Proceedings of the Eighteenth Design Automation Conference*, July 1981, pp. 786-790.

[CAS78] Case, G. R., J. D. Stauffer, "Salogs-IV A Program to Perform Logic Simulation and Fault Diagnosis," *Proceedings of the 15th Design Automation Conference*, Las Vegas, June 1978, pp. 392-397.

[CHA76] Chappell, S. G., P. R. Menon, J. F. Pellegrin, A. Schowe, "Functioinal Simulation in the Lamp System," *Proceedings of the 13th Design Automation Conference*, San Francisco, 1976, pp. 42-47.

[CHE78] Chen, A. C., J. E. Coffman, "Multi-Sim, A. Dynamic Multi-Level Simulator," *Proceedings of the 15th Design Automation Conference*,

Las Vegas, June 1978, pp. 386-391.

[CHU65] Chu, C. Y., "An ALGOL-like Computer Design Language," *Communications of the ACM*, Vol. 8, No. 10, Oct. 1965, pp. 607-615.

[CHU74] Chu, Yaohan, "Introducing CDL," *Computer*, December 1974.

[COE83] Coelho, D., and W.M.VanCleemput, "HELIX, A Tool for Multi-Level Simulation of VLSI Systems," *International Semi-Custom IC Conference*, London, November, 1983.

[COE83] Coelho, D., "HELIX, A Tool for Multi-Level Simulation of VLSI Systems," *International Semi-Custom IC Conference*,November,*1983*.

[COE84] Coelho, D., "Behavioral Simulation of LSI and VLSI Circuits," *VLSI Design Magazine*, February, 1984.

[COE84] Coelho, D., and C. Neti, "Timing Verification Using a General Behavioral Simulator," *ICCD*, October, 1984.

[COE85] Coelho, D., "Implementation Tradeoffs in High Level Simulators and How They Affect the User," *ADEE West*, March 1985.

[COE85] Coelho, D., "High--Level Design Using HELIX," ACM Computer Science Conference, March 1985

[COE85] Coelho, D., "Integrating the Electronic Development Environment with a Multi--Level Behavioral Simulator," 1985 Summer Computer Simulation Conference, July 1985

[COE85] Coelho, D., "Hardware Design Standards - A CAD Vendors Perspective," Computer Standards Conference, May 1986 future directions

[COP74] Coplaner, H. D., and J. A. Janhu, "Top Down Approach to LSI System Design." *Computer Design* Vol. 13, No. 8, Aug. 1974, pp. 143-148.

[COR79] Cory, W. E., J. R. Duley, W. M. vanCleemput, *An Introduction to the DDL-P Language*, Computer Systems Lab. Technical Report No. 163, Stanford University, Stanford, Ca. March 1979.

[COR79-2] Cory, W. E., *Syndia User's Guide* Systems Lab. Technical Report No. 176, Stanford University, Stanford, Ca. August 1979.

[COR80] Cory, W.E. and W.M.vanCleemput. "Developments in Verification of Design Correctness - A Tutorial". In Rex Fice (editor), *Tutorial - VLSI Support Technologies: Computer-Aided Design, Testing and Packaging*, pages 169-177. IEEE Computer Society Press, Los Alamitos, CA, 1982. Reprinted from Proceedings of the 17th Design Automation Conference, pages 156-164, IEEE and ACM, Minneapolis, June 1980.

[COR81], Cory, W.E., "Symbolic Simulation for Functional Verification with ADLIB and SDL," In Proceedings of the 18th Design Automation Conference, pages 82-89. IEEE and ACM, Nashville, June 1981. for chapter 5

[CRA79] Crawford, J. D., A. R. Newton, D. O. Pederson, G. R. Boyle, "A Unified Hardware Description Language for CAD Programs," *Proceedings of the 4th International Symposium on Computer Hardware Description Languages*, Palo Alto, Ca., Oct., 1979, pp. 151-154.

[DAH70] Dahl, O. J., B. Myhrhaug and K. Nygaard, *Simula Common Base Language*, Norwegian Computing Center, Oslo, Publication S-22, Oct. 1970.

[DEM79] Demou, H., G. Arnout, *DIANA - Principles of Operation*, Lisco Inc., Technical Report Number D04, Belgium 1979.

[DIE75] Dietmeyer, D. L., and J. R. Duley, "Register Transfer Languages and Their Translation," in *Digital System Design Automation*, edited by M. A. Breuer, Computer Science Press, Inc., Woodland Hills, Ca., 1975, pp. 117-218.

[DIJ75] Dijkstra, E., "Guarded Commands, Nondeterminacy and Formal Derivation of Programs," *Communications of the ACM*, 18(8), August, 1975.

[DUN83] Dunlugol, D, H.J. Derlam, P. Stivers, G.G. Schrooten, "Local Relaxation Algorithms for Event-Driven SImulation of MOS Networks Including Assignable Delay Modelling," *IEEE Transactions on CAD*, Vol CAD-2 Number 3, July, 1983.

[FAL64] Falkoff, A. D., K. E. Iverson, E. H. Sussenguth, "A Formal Description of the SYSTEM/360," *IBM Systems Journal*, Volume 3, Number 3, 1964, pp. 198-262.

[GAR77] Gardner, Robert I., "Multi-Level Modeling in SARA,"

Proceedings of the of Symposium on Design Automation and Microprocessors, Palo Alto, Feb. 24-25, 1977, pp. 63-67.

[GEH86] Gehani, N. H. and W. D. Roome., "Concurrent C," To appear in *Software--Practice & Experience*.

[GOO79] Goodel, R., "An ISPS Microassembler," *Proceedings of the 4th International Symposium on Computer Hardware Description Languages*, Palo Alto, Ca., Oct. 1979, pp. 62-67.

[HAN77] Hansen, Per Brinch, *The Architecture of Concurrent Programs*, Prentice-Hall, Englewood Cliffs, New Jersey, 1977.

[HP] Hewlett-Packard, *A Pocket Guide to Hewlett-Packard Computers*, Palo Alto, Ca.

[HIL79-1] Hill, Dwight D. and W. M. vanCleemput, "SABLE : A Tool for Generating Structured, Multi-level Simulations," *Proceedings of the 16th Design Automation Conference*, San Diego, Ca., June 1979.

[HIL79-2] Hill, D. D., *ADLIB - SABLE User's Guide*, Computer Systems Lab, Technical Report No. 177, Stanford Univ., Stanford, 1979.

[HIL79-3] Hill, D. D., "ADLIB: A Modular, Strongly-Typed Computer Design Language," *Proceedings of the 4th International Symposium on Computer Hardware Description Languages*, Palo Alto, Ca., Oct. 1979, pp. 75-81.

[HIL80] Hill, D. D., *Language and Environment for Multi-Level Simulation*, Ph. D. Thesis, Stanford University, 1980. (available through University Microfilms).

[HIL73] Hill, Frederick J. and Gerald R. Peterson, *Digital System Hardware Organization and Design*, John Wiley and Sons, 1973.

[HIL79] Hill, F. J., Z. Navabi, "Extending Second Generation AHPL Software to Accommodate AHPL III," *Proceedings of the 4th International Symposium on Computer Hardware Description Languages*, Palo Alto, Ca., Oct. 1979, pp. 47-53.

[I73] Intel, *8008 8 Bit Parallel CPU Users Manual*, Santa Clara, Ca., 1973.

[JEN74] Jensen, K. and N. Wirth, *Pascal User Manual and Report*,

Springer-Verlag, New York, 1974.

[JOH71] Johnston, John B., "The Contour Model of Block Structured Processes," *Sigplan Notices*, Feb. 1971, pp. 55.

[KIN75] King, J. C., "A New Approach to Program Testing," *ACM Sigplan Notices*, 10(6):228-233, June, 1975.

[KIN76] King, J. C., "Symbolic Execution in Program Testing," *Communications of the ACM*, 19(7):385-394, July, 1976.

[KIV69] Kiviat, P. J., R. Villanueva, and H. Markowitz, *The SIMSCRIPT II Programming Language*, Prentice Hall, Inc., Englewood Cliffs, N.J. 1969.

[KOH70] Kohavi, Zvi, *Switching and Finite Automata Theory*, McGraw-Hill Book Company, San Francisco, Ca., 1970.

[KUS76] Kusik, .R and P. Wesley, "Hierarchical Logic Simulation for Digital Systems Development," Proceedings of Electro/76, Boston, Mass,May 1976,pp 26.3.1-26.3.8.

[LAN79] Langlet, T., *private communication*, Burroughs Corporation, Mission Viejo, Ca. 1979.

[LOS75] Losleben, P., *Design Validation in Hierarchical Systems*, Proceedings of the 12th Design Automation Conference, Boston, 1975, pp. 431-438.

[MCW78] McWilliams, T. M. and L. C. Widdoes, *SCALD: Structured Computer-Aided Logic Design*, Digital Systems Lab. Tech. Report No. 152, Stanford U., March 1978.

[MAC75] MacDougall, M. H., "System Level Simulation," in *Digital System Design Automation*, edited by M. A. Breuer, Computer Science Press, Inc., Woodland Hills, 1975, pp. 35-62.

[MCW80] McWilliams, T. M., "Verification of Timing Constraints on Large Digital Systems," *Proceedings of the 17th DAC*, Minneapolis, Minn, June 1980, pp 139-147.

[MEL84] Melamed, B., R. J. T. Morris, "Visual Simulation: The Performance Analysis Workstation," *Computer*, August, 1985, pp. 87-94.

[NAG73] Nagel, L. W. and D. O. Pedersim, *SPICE (Simulatoin Program with Integrated Circuit Emphasis*, Berkeley, Calif. University of California, Electronics Research Laboratory, Memorandum ERL-M382, April 12, 1973.

[NEW78] Newton, R. A., *The Simulation of LSI Circuits*, University of Cal., Berkeley, Memo No. UCB/ERL M78/52, July 1978.

[ORG78] Organick, Elliott I., A. I. Forsythe, R. P. Plummer, *Programming Language Structures*, Academic Press, San Francisco, 1978.

[PAR79] Parker, A., D. E. Thomas, S. Crocker, R. G. G. Cattell, "ISPS: A Retrospective View," *Proceedings of the 4th International Symposium on Computer Hardware Description Languages*, Palo Alto Ca., Oct. 1979, pp. 21-27.

[ROH77] Rohmer, J. "SSH Simulateur de Systems Hierarchises," *Proceedings of the 10th Annual Simulation Symposium*, 1977 pp. 109-127.

[ROS75] Rose, C. W. and M. Albarran, "Modeling and Design Description of Hierarchical Hardware / Software Systems," *Proceedings of the 12th Design Automation Conference*, Boston, 1975, pp. 421-430.

[SCH74] Schriber, T. J., *Simulation Using GPSS*, John Wiley & Sons, New York, 1974.

[SZY72] Szygenda, S. A., "TEGAS - Anatomy of a General Purpose Test Generation and Simulation System for Digital Logic," *Proceedings of the Design Automation Workshop*, Dallas, Texas, June 1972, pp. 116-127.

[SZY77] Szygenda, S. A., A. A. Lekkos, "Integrated Techniques for Functional and Gate-Level Digital Logic Simulation," *Proceedings of the 10th Design Automation Workshop*, Portland, Oregon, June 1973, pp. 159-172.

[SZY84] Szymanski, T., *private communications*, AT&T Bell Laboratories, 1984.

[TAN81] Tanenbaum, A., *Computer Networks*, Prentice Hall, 1981.

[TI73] Texas Instruments, *The TTL Data Book for Design Engineers*, Dallas, Texas 1973.

[TOK78] Tokoro, M., M. Sato, M. Ishigami, E. Tamura, T. Ishimitsu, H. Ohara, *A Module Level Simulation Technique for Systems Composed of LSI's and MSI's*, Proceedings of the 15th Design Automation Conference, Las Vegas 1978, pp. 418-427.

[VAN77] vanCleemput, W. M., "An Hierarchical Language for the Structural Description of Digital Systems," *Proceedings of the 14th Design Automation Conference*, New Orleans, 1977, pp. 377-385.

[VHD85] *VHDL Language Reference Manual, Version 7.2*, Intermetrics, Inc, Bethesda, MD IR-MD-045-2

[WAL79] Wallace, J. J., A. C. Parker, "SLIDE: An I/O Hardware Descriptive Language," *Proceedings of the 4th International Symposium on Computer Hardware Description Languages*, Palo Alto, Ca., Oct. 1979, pp. 82-88.

[WIL76] Wilcox, P., and H. Rombeek, "F/LOGIC - An Interactive Fault and Logic Simulator for Digital Circuits,"

[WIR75] Wirth, N., "An Assessment of the Programming Language Pascal," *IEEE Transactions of Software Engineering*, Vol. SE 1, 2, June 1975.

[WIR78] Wirth, N., *Modula - 2*, Instut fur Informatik ETH CH-8092, Zurich, Dec. 1978. *Proceedings of the 13th Design Automation Conference*, San Francisco, 1976, pp. 68-73.

[WOL78] Wold, M.A. "Design Verification and Performance Analysis," *Proceedings of the Fifteenth Design Automation Conference*, Las Vegas,Nevada, June 1978,pp 264-270.

[ZIM80] Zimmerman, H., "OSI Reference Model, the ISO Model of the Architecture for Open Systems Interconnectsion, *IEEE Transactions on Communications*, vol COM-28, pp 425-432, April 1980.

APPENDIX F

INDEX